**Oxford Applied Mathematics
and Computing Science Series**

T0177781

Oxford Applied Mathematics and Computing Science Series

RAYMOND HILL
University of Salford

A First Course in Coding Theory

CLARENDON PRESS • OXFORD

OXFORD

UNIVERSITY PRESS

Great Clarendon Street, Oxford OX2 6DP
Oxford University Press is a department of the University of Oxford.
It furthers the University's objective of excellence in research, scholarship,
and education by publishing worldwide in
Oxford New York
Auckland Cape Town Dar es Salaam Hong Kong Karachi
Kuala Lumpur Madrid Melbourne Mexico City Nairobi
New Delhi Shanghai Taipei Toronto
With offices in
Argentina Austria Brazil Chile Czech Republic France Greece
Guatemala Hungary Italy Japan South Korea Poland Portugal
Singapore Switzerland Thailand Turkey Ukraine Vietnam

Oxford is a registered trade mark of Oxford University Press
in the UK and in certain other countries
Published in the United States
by Oxford University Press Inc., New York

ISBN 978-0-19-853803-5

Printed in the United Kingdom by
Lightning Source UK Ltd., Milton Keynes

To
Susan, Jonathan, and Kathleen

Preface

The birth of coding theory was inspired by a classic paper of Shannon in 1948. Since then a great deal of research has been devoted to finding efficient schemes by which digital information can be coded for reliable transmission through a noisy channel. Error-correcting codes are now widely used in applications such as returning pictures from deep space, design of registration numbers, and storage of data on magnetic tape. Coding theory is also of great mathematical interest, relying largely on ideas from pure mathematics and, in particular, illustrating the power and the beauty of algebra. Several excellent textbooks have appeared in recent years, mostly at graduate level and assuming a fairly advanced level of mathematical knowledge or sophistication. Yet the basic ideas and much of the theory of coding are readily accessible to anyone with a minimal mathematical background. (For a recent article advocating the inclusion of algebraic coding theory in the undergraduate curriculum, see Brinn (1984).)

The aim of this book is to provide an elementary treatment of the theory of error-correcting codes, assuming no more than high school mathematics and the ability to carry out matrix arithmetic. The book is intended to serve as a self-contained course for second or third year mathematics undergraduates, or as a readable introduction to the mathematical aspects of coding for students in engineering or computer science.

The first eight chapters comprise an introductory course which I have taught as part of second year undergraduate courses in discrete mathematics and in algebra. (There is much to be said for teaching coding theory immediately after, or concurrently with, a course in algebra, for it reinforces with concrete examples many of the ideas involved in linear algebra and in elementary group theory.) I have also used the text as a whole as a Master's course taken by students whose first degree is not necessarily in mathematics. The last eight chapters are largely independent of one another and so courses can be varied to suit requirements. For example, Chapters 9, 10, 14, and 15 might be omitted by students who are not specialist mathematicians.

The book is concerned almost exclusively with block codes for correcting random errors, although the last chapter includes a brief discussion of some other codes, such as variable length source codes and cryptographic codes. The treatment throughout is motivated by two central themes: the problem of finding the best codes, and the problem of decoding such codes efficiently.

One departure from several standard texts is that attention is by no means restricted to binary codes. Indeed, consideration of codes over fields of order a prime number enables much of the theory, including the construction and decoding of BCH codes, to be covered in an elementary way, without needing to work with the rather more complex fields of order 2^h $(h > 1)$.

Another feature is the large number of exercises, at varying levels of difficulty, at the end of each chapter. The inclusion of the solutions at the end makes the book suitable for self-learning or for use as a reading course. I believe that the best way to understand a subject is by solving problems and so the reader is urged to make good attempts at the exercises before consulting the solutions.

Finally, it is hoped that the reader will be given a taste for this fascinating subject and so encouraged to read the more advanced texts. Outstanding amongst these is MacWilliams and Sloane (1977); the size of its bibliography—nearly 1500 articles—is a measure of how coding theory has grown since 1948. Also highly recommended are Berlekamp (1968), Blahut (1983), Blake and Mullin (1976), Cameron and van Lint (1980), Lin and Costello (1983), van Lint (1982), McEliece (1977), Peterson and Weldon (1972), and Pless (1982).

Salford
February 1985

Acknowledgements

I am grateful to Professors P. G. Farrell and J. H. van Lint, and Drs J. W. P. Hirschfeld, R. W. Irving, L. O'Carroll, and R. Sandling for helpful comments and suggestions, and to B. Banieqbal for acquainting me with the work of Ramanujan, which now features prominently in Chapter 11.

I should also like to thank Susan Sharples for her excellent typing of the manuscript.

R.H.

Contents

Notation

For the reader who is unfamiliar with the notation of modern set theory, we introduce below all that is required in this book.

A *set* is simply a collection of objects. In this book we shall make use of the following sets (among others):

R: the set of real numbers.

Z: the set of integers (positive, negative, or zero).

Z_n: the set of integers from 0 to $n - 1$ inclusive.

The objects in a set are often called its *elements* or its *members*. If x is an element of the set S, we write $x \in S$, which is read 'x belongs to S' or 'x belonging to S' as the context requires. If x is not an element of S we write $x \notin S$. Thus $2 \in Z$ but $\frac{1}{2} \notin Z$. Two sets are *equal* if they contain precisely the same elements. The set consisting precisely of elements $x_1, x_2, \ldots, x_n$ is often denoted by $\{x_1, x_2, \ldots, x_n\}$. For example, $Z_3 = \{0, 1, 2\}$. Also $Z_3 = \{0, 2, 1\} = \{2, 1, 0\}$.

If S is a set and P a property (or combination of properties) which elements x of S may or may not possess, we can define a new set with the notation

$$\{x \in S \mid P(x)\}$$

which denotes 'the set of all elements belonging to S which have property P'. For example, the set of positive integers could be written $\{x \in Z \mid x > 0\}$ which we read as 'the set of elements x belonging to Z such that x is greater than 0'. The set of all even integers can be denoted by $\{2n \mid n \in Z\}$.

A set T is called a *subset* of a set S if all the elements of T belong to S. We then say that 'T is contained in S' and write $T \subseteq S$, or that 'S contains T' and write $S \supseteq T$.

If S and T are sets we define the *union* $S \cup T$ of S and T to be the set of all elements in either S or T. We define the *intersection* $S \cap T$ of S and T to be the set of all elements which are members of both S and T. Thus

$$S \cup T = \{x \mid x \in S \text{ or } x \in T\},$$
$$S \cap T = \{x \mid x \in S \text{ and } x \in T\}$$

If S and T have no members in common, we say that S and T are *disjoint*.

The *order* or *cardinality* of a finite set S is the number of elements in S and is denoted by $|S|$. For example, $|Z_n| = n$.

Given sets S and T we denote by (s, t) an *ordered pair* of elements where $s \in S$ and $t \in T$. Two ordered pairs (s_1, t_1) and (s_2, t_2) are defined to be *equal* if and only if $s_1 = s_2$ and $t_1 = t_2$. Thus if $S = T = Z$, $(0, 1) \neq (1, 0)$. The *Cartesian product* of S and T, denoted by $S \times T$, is defined to be the set of all ordered pairs (s, t) such that $s \in S$ and $t \in T$. The product $S \times S$ is denoted by S^2. Thus

$$S^2 = \{(s_1, s_2) \mid s_1 \in S, s_2 \in S\}.$$

If S and T are finite sets, then

$$|S \times T| = |S| \cdot |T|$$

for, in forming an element (s, t) of $S \times T$, we have $|S|$ choices for s and $|T|$ choices for t. In particular $|S^2| = |S|^2$.

More generally we define the *Cartesian product* of n sets $S_1, S_2, \ldots, S_n$ to be a set of *ordered n-tuples* thus:

$$S_1 \times S_2 \times \cdots \times S_n = \{(s_1, s_2, \ldots, s_n) \mid s_i \in S_i, i = 1, 2, \ldots, n\}.$$

Two ordered n-tuples $(s_1, s_2, \ldots, s_n)$ and $(t_1, t_2, \ldots, t_n)$ are defined to be equal if and only if $s_i = t_i$ for $i = 1, 2, \ldots, n$. If $S_1 = S_2 = \cdots = S_n = S$, the product is denoted by S^n. For example,

$$R^3 = \{(x, y, z) \mid x \in R, y \in R, z \in R\}$$

is a set-theoretic description of coordinatized 3-space. If S is finite, then clearly

$$|S^n| = |S|^n.$$

Finally we remark that in this book we shall often write an ordered n-tuple $(x_1, x_2, \ldots, x_n)$ simply as $x_1 x_2 \cdots x_n$.

1 Introduction to error-correcting codes

Error-correcting codes are used to correct errors when messages are transmitted through a noisy communication channel. For example, we may wish to send binary data (a stream of 0s and 1s) through a noisy channel as quickly and as reliably as possible. The channel may be a telephone line, a high frequency radio link, or a satellite communication link. The noise may be human error, lightning, thermal noise, imperfections in equipment, etc., and may result in errors so that the data received is different from that sent. The object of an error-correcting code is to encode the data, by adding a certain amount of redundancy to the message, so that the original message can be recovered if (not too many) errors have occurred. A general digital communication system is shown in Fig. 1.1. The same model can be used to describe an information storage system if the storage medium is regarded as a channel; a typical example is a magnetic-tape unit including writing and reading heads.

Figure 1.1

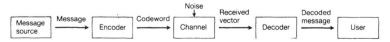

Let us look at a very simple example in which the only messages we wish to send are 'YES' and 'NO'.

Example 1.2

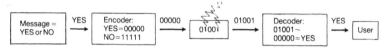

Here two errors have occurred and the decoder has decoded the received vector 01001 as the 'nearest' codeword which is 00000 or YES.

A *binary code* is just a given set of sequences of 0s and 1s which are called *codewords*. The code of Example 1.2 is {00000, 11111}. If the messages YES and NO are identified with the symbols 0 and 1 respectively, then each message symbol is encoded simply by repeating the symbol five times. The code is called a *binary repetition code of length 5*. This is an example of how 'redundancy' can be added to messages to protect them against noise. The extra symbols sent are themselves subject to error and so there is no way to *guarantee* accuracy; we just try to make the probability of accuracy as high as possible. Clearly, a good code is one in which the codewords have little resemblance to each other.

More generally, a *q-ary code* is a given set of sequences of symbols where each symbol is chosen from a set $F_q = \{\lambda_1, \lambda_2, \dots, \lambda_q\}$ of q distinct elements. The set F_q is called the *alphabet* and is often taken to be the set $Z_q = \{0, 1, 2, \dots, q-1\}$. However, if q is a prime power (i.e. $q = p^h$ for some prime number p and some positive integer h) then we often take the alphabet F_q to be the finite field of order q (see Chapter 3). As we have already seen, 2-ary codes are called *binary codes*; 3-ary codes are sometimes referred to as *ternary codes*.

Example 1.3   (i)   The set of all words in the English language is a code over the 26-letter alphabet {A, B, . . . , Z}.

(ii)   The set of all street names in the city of Salford is a 27-ary code (the space between words is the 27th symbol) and provides a good example of poor encoding, for two street names on the same estate are HILLFIELD DRIVE and MILLFIELD DRIVE.

A code in which each codeword is a sequence consisting of a fixed number n of symbols is called a *block code* of *length n*. From now on we shall restrict our attention almost exclusively to such codes and so by 'code' we shall always mean 'block code'.

A code C with M codewords of length n is often written as an $M \times n$ array whose rows are the codewords of C. For example, the binary repetition code of length 3 is

$$000$$
$$111.$$

Let $(F_q)^n$ denote the set of all ordered n-tuples $\mathbf{a} = a_1 a_2 \cdots a_n$ where each $a_i \in F_q$. The elements of $(F_q)^n$ are called *vectors* or *words*. The order of the set $(F_q)^n$ is q^n. A *q-ary code of length n* is just a subset of $(F_q)^n$.

Example 1.4 The set of all 10-digit telephone numbers in the United Kingdom is a 10-ary code of length 10. Little thought appears to have been given to allocating numbers so that the frequency of 'wrong numbers' is minimized. Yet it is possible to use a code of over 82 million 10-digit telephone numbers (enough for the needs of the UK) such that if just one digit of any number is misdialled the correct connection can nevertheless be made. We will construct this code in Chapter 7 (Example 7.12).

Example 1.5 Suppose that HQ and X have identical maps gridded as shown in Fig. 1.6 but that only HQ knows the route indicated, avoiding enemy territory, by which X can return safely to HQ. HQ can transmit binary data to X and wishes to send the route NNWNNWWSSWWNNNNWWN. This is a situation where reliability is more important than speed of transmission. Consider how the four messages N, S, E, W can be encoded into binary codewords. The fastest (i.e. shortest) code we could use is

$$C_1 = \begin{cases} 0\ 0 = N \\ 0\ 1 = W \\ 1\ 0 = E \\ 1\ 1 = S. \end{cases}$$

Figure 1.6

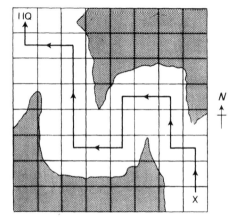

That is, we identify the four messages N, W, E, S with the four vectors of $(F_2)^2$. Let us see how, as in Example 1.1, redundancy can be added to protect these message vectors against noise. Consider the length 3 code C_2 obtained by adding an extra digit as follows.

$$C_2 = \begin{cases} 0\ 0\ 0 \\ 0\ 1\ 1 \\ 1\ 0\ 1 \\ 1\ 1\ 0. \end{cases}$$

This takes longer than C_1 to transmit but if there is any single error in a codeword, the received vector cannot be a codeword (check this!) and so the receiver will recognize that an error has occurred and may be able to ask for the message to be retransmitted. Thus C_2 has the facility to *detect* any single error; we say it is a single-error-detecting code.

Now suppose X can receive data from HQ but is unable to seek retransmission, i.e. we have a strictly *one-way channel*. A similar situation might well apply in receiving photographs from deep space or in the playing back of an old magnetic tape, and in such cases it is essential to extract as much information as possible from the received vectors. By suitable addition of two further digits to each codeword of C_2 we get the length 5 code

$$C_3 = \begin{cases} 0\ 0\ 0\ 0\ 0 \\ 0\ 1\ 1\ 0\ 1 \\ 1\ 0\ 1\ 1\ 0 \\ 1\ 1\ 0\ 1\ 1. \end{cases}$$

If a single error occurs in any codeword of C_3, we are able not only to detect it but actually to *correct* it, since the received vector will still be 'closer' to the transmitted codeword than to any other. (Check that this is so and also that if used only for error-detection C_3 is a two-error-detecting code).

We have so far talked rather loosely about a vector being 'closer' to one codeword than to another and we now make this concept precise by introducing a distance function on $(F_q)^n$, called the Hamming distance.

The *(Hamming) distance* between two vectors $\mathbf{x}$ and $\mathbf{y}$ of $(F_q)^n$ is the number of places in which they differ. It is denoted by $d(\mathbf{x}, \mathbf{y})$. For example, in $(F_2)^5$ we have $d(00111, 11001) = 4$, while in $(F_3)^4$ we have $d(0122, 1220) = 3$.

The Hamming distance is a legitimate distance function, or metric, since it satisfies the three conditions:

(i) $d(\mathbf{x}, \mathbf{y}) = 0$ if and only if $\mathbf{x} = \mathbf{y}$.
(ii) $d(\mathbf{x}, \mathbf{y}) = d(\mathbf{y}, \mathbf{x})$ for all $\mathbf{x}, \mathbf{y} \in (F_q)^n$.
(iii) $d(\mathbf{x}, \mathbf{y}) \le d(\mathbf{x}, \mathbf{z}) + d(\mathbf{z}, \mathbf{y})$ for all $\mathbf{x}, \mathbf{y}, \mathbf{z} \in (F_q)^n$.

The first two conditions are very easy to verify. The third, known as the *triangle inequality*, is verified as follows. Note that $d(\mathbf{x}, \mathbf{y})$ is the minimum number of changes of digits required to change $\mathbf{x}$ to $\mathbf{y}$. But we can also change $\mathbf{x}$ to $\mathbf{y}$ by first making $d(\mathbf{x}, \mathbf{z})$ changes (changing $\mathbf{x}$ to $\mathbf{z}$) and then $d(\mathbf{z}, \mathbf{y})$ changes (changing $\mathbf{z}$ to $\mathbf{y}$). Thus $d(\mathbf{x}, \mathbf{y}) \le d(\mathbf{x}, \mathbf{z}) + d(\mathbf{z}, \mathbf{y})$.

The Hamming distance will be the only metric considered in this book. However, it is not the only one possible and indeed may not always be the most appropriate. For example, in $(F_{10})^3$ we have $d(428, 438) = d(428, 468)$, whereas in practice, e.g. in dialling a telephone number, it might be more sensible to use a metric in which 428 is closer to 438 than it is to 468.

Let us now consider the problem of decoding. Suppose a codeword $\mathbf{x}$, unknown to us, has been transmitted and that we receive the vector $\mathbf{y}$ which may have been distorted by noise. It seems reasonable to decode $\mathbf{y}$ as that codeword $\mathbf{x}'$, hopefully $\mathbf{x}$, such that $d(\mathbf{x}', \mathbf{y})$ is as small as possible. This is called *nearest neighbour decoding*. This strategy will certainly maximize the decoder's likelihood of correcting errors provided the following assumptions are made about the channel.

(i)   Each symbol transmitted has the same probability $p(<\tfrac{1}{2})$ of being received in error.

(ii) If a symbol is received in error, then each of the $q - 1$ possible errors is equally likely.

Such a channel is called a *q-ary symmetric channel*. The binary symmetric channel is shown in Fig. 1.7.

Figure 1.7

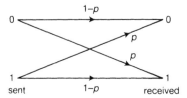

p is called the *symbol error probability* of the channel.

If the binary symmetric channel is assumed and if a particular binary codeword of length n is transmitted, then the probability that no errors will occur is $(1-p)^n$, since each symbol has probability $(1-p)$ of being received correctly. The probability that one error will occur in a specified position is $p(1-p)^{n-1}$. The probability that the received vector has errors in precisely i specified positions is $p^i(1-p)^{n-i}$. Since $p < \frac{1}{2}$, the received vector with no errors is more likely than any other; any received vector with one error is more likely than any with two or more errors, and so on. This confirms that, for a binary symmetric channel, nearest neighbour decoding is also *maximum likelihood decoding*.

Example 1.8   Consider the binary repetition code of length 3

$$C = \begin{cases} 000 \\ 111 \end{cases}.$$

Suppose the codeword 000 is transmitted. Then the received vectors which will be decoded as 000 are 000, 100, 010 and 001. Thus the probability that the received vector is decoded as the transmitted codeword 000 is

$$(1-p)^3 + 3p(1-p)^2 = (1-p)^2(1+2p).$$

Note that, by symmetry, the probability is the same if the transmitted codeword is 111. Thus we can say that the code C has a *word error probability*, denoted by $P_{\mathrm{err}}(C)$, which is independent of the codeword transmitted. In this example, we have
$$P_{\mathrm{err}}(C) = 1 - (1-p)^2(1+2p) = 3p^2 - 2p^3.$$

In order to compare probabilities given by such polynomials in p, it is useful to assign an appropriate numerical value to p. For example we might assume that, on average, the channel causes one symbol in a hundred to be received in error, i.e. $p = 0.01$. In this case $P_{err}(C) = 0.000\,298$ and so approximately only one word in 3355 will reach the user in error.

We will show in Chapter 6 that a very important class of codes, called linear codes, all have the property that the word error probability is independent of the actual codeword sent. For a general code, a brute-force decoding scheme is to compare the received vector with all codewords and to decode as the nearest. This is impractical for large codes and one of the aims of coding theory is to find codes which can be decoded by faster methods than this. We shall see in Chapters 6 and 7 that linear codes have elegant decoding schemes.

An important parameter of a code C, giving a measure of how good it is at error-correcting, is the *minimum distance*, denoted $d(C)$, which is defined to be the smallest of the distances between distinct codewords. That is,

$$d(C) = \min\,\{d(\mathbf{x}, \mathbf{y}) \mid \mathbf{x}, \mathbf{y} \in C, \mathbf{x} \neq \mathbf{y}\}.$$

For example, it is easily checked that for the codes of Example 1.5, $d(C_1) = 1$, $d(C_2) = 2$ and $d(C_3) = 3$.

Theorem 1.9 (i) A code C can detect up to s errors in any codeword if $d(C) \geqslant s + 1$.

(ii) A code C can correct up to t errors in any codeword if $d(C) \geqslant 2t + 1$.

Proof (i) Suppose $d(C) \geqslant s + 1$. Suppose a codeword $\mathbf{x}$ is transmitted and s or fewer errors are introduced. Then the received vector cannot be a different codeword and so the errors can be detected.

(ii) Suppose $d(C) \geqslant 2t + 1$. Suppose a codeword $\mathbf{x}$ is transmitted and the vector $\mathbf{y}$ received in which t or fewer errors have occurred, so that $d(\mathbf{x}, \mathbf{y}) \leqslant t$. If $\mathbf{x}'$ is any codeword other than $\mathbf{x}$, then $d(\mathbf{x}', \mathbf{y}) \geqslant t + 1$. For otherwise, $d(\mathbf{x}', \mathbf{y}) \leqslant t$, which implies, by the triangle inequality, that $d(\mathbf{x}, \mathbf{x}') \leqslant d(\mathbf{x}, \mathbf{y}) + d(\mathbf{x}', \mathbf{y}) \leqslant 2t$, contradicting $d(C) \geqslant 2t + 1$. So $\mathbf{x}$ is the nearest codeword to $\mathbf{y}$ and nearest neighbour decoding corrects the errors.

[*Note*: The reader may find Remark 2.12 helpful in clarifying this proof.]

Corollary 1.10 If a code C has minimum distance d, then C can be used either (i) to detect up to $d - 1$ errors, or (ii) to correct up to $\lfloor (d - 1)/2 \rfloor$ errors in any codeword.
($\lfloor x \rfloor$ denotes the greatest integer less than or equal to x).

Proof (i) $d \geqslant s + 1$ iff $s \leqslant d - 1$. (ii) $d \geqslant 2t + 1$ iff $t \leqslant (d - 1)/2$.

For example, if $d(C) = 3$, then C can be used either as a single-error-correcting code or as a double-error-detecting code. More generally we have:

$d(C)$	Number of errors detected by C	Number of errors corrected by C
1	0	0
2	1	0
3	2	1
4	3	1
5	4	2
6	5	2
7	6	3
.	.	.
.	.	.
.	.	.

The following notation will be used extensively and should be memorized.

An (n, M, d)-*code* is a code of length n, containing M codewords and having minimum distance d.

Examples 1.11 (i) In Example 1.5, C_1 is a $(2, 4, 1)$-code, C_2 a $(3, 4, 2)$-code and C_3 a $(5, 4, 3)$-code.
(ii) The q-*ary repetition code* of length n whose codewords are

$$
\begin{array}{ccc}
0 & 0 & \cdots & 0 \\
1 & 1 & \cdots & 1 \\
\vdots & \vdots & & \vdots \\
(q - 1) & (q - 1) & \cdots & (q - 1)
\end{array}
$$

is an (n, q, n)-code.

Example 1.12 The code used by Mariner 9 to transmit pictures from Mars was a binary (32, 64, 16)-code, called a Reed–Muller code. This code, which will be constructed in Exercise 2.19, is well suited to very noisy channels and also has a fast decoding algorithm. How the code was used will be described in the following brief history of the transmission of photographs from NASA space probes.

The transmission of photographs from deep-space

1965: Mariner 4 was the first spaceship to photograph another planet, taking 22 complete photographs of Mars. Each picture was broken down into 200×200 picture elements. Each element was assigned a binary 6-tuple representing one of 64 brightness levels from white ($=000000$) to black ($=111111$). Thus the total number of bits (i.e. binary digits) per picture was 240 000. Data was transmitted at the rate of $8\frac{1}{3}$ bits per second and so it took 8 hours to transmit a single picture!

1969–1972: Much improved pictures of Mars were obtained by Mariners 6, 7 and 9 (Mariner 8 was lost during launching). There were three important reasons for this improvement:

(1) Each picture was broken down into 700×832 elements (cf. 200×200 of Mariner 4 and 400×525 of US commercial television).

(2) Mariner 9 was the first spaceship to be put into orbit around Mars.

(3) The powerful Reed–Muller (32, 64, 16)-code was used for error correction. Thus a binary 6-tuple representing the brightness of a dot in the picture was now encoded as a binary codeword of length 32 (having 26 redundant bits).

The data transmission rate was increased from $8\frac{1}{3}$ to 16 200 bits per second. Even so, picture bits were produced by Mariner's cameras at more than 100 000 per second, and so data had to be stored on magnetic tape before transmission.

1976: Viking 1 landed softly on Mars and returned high-quality *colour* photographs.

Surprisingly, transmission of a colour picture in the form of binary data is almost as easy as transmission of a black-and-white one. It is achieved simply by taking the same black-and-white photograph several times, each time through a different coloured filter. The black-and-white pictures are then transmitted as already described and the colour picture reconstructed back on Earth.

5 March 1979: High-resolution colour pictures of Jupiter and its moons were returned by Voyager 1.

12 November 1980: Voyager 1 returned the first high-resolution pictures of Saturn and its moons.

25 August 1981: Voyager 2 returned further excellent pictures of Saturn.

And to come:

24 January 1986: Voyager 2 passes Uranus.

24 August 1989: Voyager 2 passes Neptune.

Exercises 1

1.1 If the following message were received from outer space, why might it be conjectured that it was sent by a race of human-like beings who have one arm twice as long as the other? [*Hint:* The number of digits in the message is the product of two prime numbers.]

0011000001100011111111011001001100100110010111110001 00100010010001001001100110

1.2 Suppose the binary repetition code of length 5 is used for a binary symmetric channel which has symbol error probability p. Show that the word error probability of the code is $10p^3 - 15p^4 + 6p^5$.

1.3 Show that a code having minimum distance 4 can be used simultaneously to correct single errors and detect double errors.

1.4 The code used by Mariner 9 will correct any received 32-tuple provided not more than . . . (how many?) errors have occurred.

1.5 (i) Show that a 3-ary $(3, M, 2)$-code must have $M \leq 9$.
    (ii) Show that a 3-ary $(3, 9, 2)$-code does exist.
    (iii) Generalize the results of (i) and (ii) to q-ary $(3, M, 2)$-codes, for any integer $q \geq 2$.

2 The main coding theory problem

A good (n, M, d)-code has small n (for fast transmission of messages), large M (to enable transmission of a wide variety of messages) and large d (to correct many errors). These are conflicting aims and what is often referred to as the 'main coding theory problem' is to optimize one of the parameters n, M, d for given values of the other two. The usual version of the problem is to find the largest code of given length and given minimum distance. We denote by $A_q(n, d)$ the largest value of M such that there exists a q-ary (n, M, d)-code.

The problem is easily solved for $d = 1$ and $d = n$, for all q:

Theorem 2.1 (i) $A_q(n, 1) = q^n$. (ii) $A_q(n, n) = q$.

Proof (i) For the minimum distance of a code to be at least 1 we require that the codewords are distinct, and so the largest q-ary $(n, M, 1)$-code is the whole of $(F_q)^n$, with $M = q^n$.

(ii) Suppose C is a q-ary (n, M, n)-code. Then any two distinct codewords of C differ in all n positions. Thus the symbols appearing in any fixed position, e.g. the first, in the M codewords must be distinct, giving $M \leqslant q$. Thus $A_q(n, n) \leqslant q$. On the other hand, the q-ary repetition code of length n (see Example 1.11(ii)) is an (n, q, n)-code and so $A_q(n, n) = q$.

Example 2.2 We will determine the value $A_2(5, 3)$. The code C_3 of Example 1.5 is a binary $(5, 4, 3)$-code and so $A_2(5, 3) \geqslant 4$. But can we do better? To show whether there exists a binary $(5, 5, 3)$-code a brute-force method would be to consider all subsets of order 5 in $(F_2)^5$ and find the minimum distance of each. Unfortunately there are over 200 000 such subsets (see Example 2.11(iii)), but, by using the following notion of equivalence, the search can be considerably reduced. We will return to Example 2.2 shortly.

Equivalence of codes

A *permutation* of a set $S = \{x_1, x_2, \ldots, x_n\}$ is a one-to-one mapping from S to itself. We denote a permutation f by

$$
\begin{pmatrix}
x_1 & x_2 & \cdots & x_n \\
\downarrow & \downarrow & & \downarrow \\
f(x_1) & f(x_2) & \cdots & f(x_n)
\end{pmatrix}.
$$

Definition Two q-ary codes are called *equivalent* if one can be obtained from the other by a combination of operations of the following types:

(A) permutation of the positions of the code;
(B) permutation of the symbols appearing in a fixed position.

If a code is displayed as an $M \times n$ matrix whose rows are the codewords, then an operation of type (A) corresponds to a permutation, or rearrangement, of the columns of the matrix, while an operation of type (B) corresponds to a re-labelling of the symbols appearing in a given column.

Clearly the distances between codewords are unchanged by such operations and so equivalent codes have the same parameters (n, M, d) and will correct the same number of errors. Indeed, under the assumptions of a q-ary symmetric channel, the performances of equivalent codes will be identical in terms of probabilities of error correction.

Examples (i) The binary code

$$
C = \begin{cases}
0\ 0\ 1\ 0\ 0 \\
0\ 0\ 0\ 1\ 1 \\
1\ 1\ 1\ 1\ 1 \\
1\ 1\ 0\ 0\ 0
\end{cases}
$$

is equivalent to the code C_3 of Example 1.5. (Apply the permutation

$$
\begin{pmatrix}
0 & 1 \\
\downarrow & \downarrow \\
1 & 0
\end{pmatrix}
$$

to the symbols in the third position of C and then interchange positions 2 and 4. Note that the codewords will be listed in a different order from that in Example 1.5).

(ii) The ternary code

$$C = \begin{cases} 0\ 1\ 2 \\ 1\ 2\ 0 \\ 2\ 0\ 1 \end{cases}$$

is equivalent to the ternary repetition code of length 3. Applying the permutation

$$\begin{pmatrix} 0 & 1 & 2 \\ \downarrow & \downarrow & \downarrow \\ 2 & 0 & 1 \end{pmatrix}$$

to the symbols in the second position and

$$\begin{pmatrix} 0 & 1 & 2 \\ \downarrow & \downarrow & \downarrow \\ 1 & 2 & 0 \end{pmatrix}$$

to the symbols in the third position of C gives the code

$$\begin{cases} 0\ 0\ 0 \\ 1\ 1\ 1. \\ 2\ 2\ 2 \end{cases}$$

Lemma 2.3 Any q-ary (n, M, d)-code over an alphabet $\{0, 1, \ldots, q-1\}$ is equivalent to an (n, M, d)-code which contains the all-zero vector $\mathbf{0} - 0\,0\,\cdots\,0$.

Proof Choose any codeword $x_1x_2\cdots x_n$ and for each $x_i \neq 0$ apply the permutation

$$\begin{pmatrix} 0 & x_i & j \\ \downarrow & \downarrow & \downarrow & \text{for all } j \neq 0, x_i \\ x_i & 0 & j \end{pmatrix}$$

to the symbols in position i.

Example 2.2 (continued) We will show not only that a binary $(5, M, 3)$-code must have $M \leq 4$ but also that the $(5, 4, 3)$-code is unique, up to equivalence.

Let C be a $(5, M, 3)$-code with $M \geq 4$. Then by Lemma 2.3 we may assume that C contains the vector $\mathbf{0} = 00000$, (replacing C by an equivalent code which does contain $\mathbf{0}$, if necessary). Now C contains at most one codeword having 4 or 5 1s, for if there were two such codewords, $\mathbf{x}$ and $\mathbf{y}$ say, then $\mathbf{x}$ and $\mathbf{y}$ would have at least 3 1s in common positions, giving $d(\mathbf{x}, \mathbf{y}) \leq 2$ and contradicting $d(C) = 3$.

Since $\mathbf{0} \in C$, there can be no codewords containing just one or two 1s and so, since $M \geq 4$, there must be at least two codewords containing exactly 3 1s. By rearranging the positions, if necessary, we may thus assume that C contains the codewords

$$0\ 0\ 0\ 0\ 0$$
$$1\ 1\ 1\ 0\ 0.$$
$$0\ 0\ 1\ 1\ 1$$

It is now very easy to show by trial and error that the only possible further codeword can be 11011.

We have thus shown that $A_2(5, 3) = 4$ and that the code which achieves this value is, up to equivalence, unique.

Restricting our attention for the time being to binary codes, we list in Table 2.4 the known non-trivial values of $A_2(n, d)$ for $n \leq 16$ and $d \leq 7$. This is taken from the table on P. 156 of Sloane (1982) which in turn is an updating of the table on P. 674 of

Table 2.4

n	$d = 3$	$d = 5$	$d = 7$
5	4	2	—
6	8	2	—
7	16	2	2
8	20	4	2
9	40	6	2
10	72–79	12	2
11	144–158	24	4
12	256	32	4
13	512	64	8
14	1024	128	16
15	2048	256	32
16	2560–3276	256–340	36–37

MacWilliams and Sloane (1977). Where the value of $A_2(n, d)$ is not known, the best available bounds are given; for example, the entry 72–79 indicates that $72 \leqslant A_2(10, 3) \leqslant 79$.

Many of the entries of Table 2.4 will be established during the course of this book (we have already verified the first entry in Example 2.2). In Chapter 16 we shall again consider Table 2.4 and review the progress we have made.

The reason why only odd values of d need to be considered in the table is that if d is an even number, then $A_2(n, d) = A_2(n - 1, d - 1)$, a result (Corollary 2.8) towards which we now proceed.

Taking F_2 to be the set $\{0, 1\}$, we define two operations on $(F_2)^n$. Let $\mathbf{x} = x_1 x_2 \cdots x_n$ and $\mathbf{y} = y_1 y_2 \cdots y_n$ be two vectors in $(F_2)^n$. Then the *sum* $\mathbf{x} + \mathbf{y}$ is the vector in $(F_2)^n$ defined by

$$\mathbf{x} + \mathbf{y} = (x_1 + y_1, x_2 + y_2, \ldots, x_n + y_n),$$

while the *intersection* $\mathbf{x} \cap \mathbf{y}$ is the vector in $(F_2)^n$ defined by

$$\mathbf{x} \cap \mathbf{y} = (x_1 y_1, x_2 y_2, \ldots, x_n y_n).$$

The terms $x_i + y_i$ and $x_i y_i$ are calculated modulo 2 (without carrying); that is, according to the addition and multiplication tables

+	0	1		·	0	1
0	0	1		0	0	0
1	1	0		1	0	1

For example $11100 + 00111 = 11011$

and $11100 \cap 00111 = 00100.$

The *weight* of a vector $\mathbf{x}$ in $(F_2)^n$, denoted $w(\mathbf{x})$, is defined to be the number of 1s appearing in $\mathbf{x}$.

Lemma 2.5   If $\mathbf{x}$ and $\mathbf{y} \in (F_2)^n$, then $d(\mathbf{x}, \mathbf{y}) = w(\mathbf{x} + \mathbf{y})$.

Proof   The sum $\mathbf{x} + \mathbf{y}$ has a 1 where $\mathbf{x}$ and $\mathbf{y}$ differ and a 0 where $\mathbf{x}$ and $\mathbf{y}$ agree.

Lemma 2.6   If $\mathbf{x}$ and $\mathbf{y} \in (F_2)^n$, then

$$d(\mathbf{x}, \mathbf{y}) = w(\mathbf{x}) + w(\mathbf{y}) - 2w(\mathbf{x} \cap \mathbf{y}).$$

Proof $d(\mathbf{x}, \mathbf{y}) = w(\mathbf{x} + \mathbf{y}) = $ (number of 1s in $\mathbf{x}$) + (number of 1s

in $\mathbf{y}$) $-$ 2(number of positions where both $\mathbf{x}$ and $\mathbf{y}$ have a
1) $= w(\mathbf{x}) + w(\mathbf{y}) - 2w(\mathbf{x} \cap \mathbf{y})$.

Theorem 2.7 Suppose d is odd. Then a binary (n, M, d)-code
exists if and only if a binary $(n + 1, M, d + 1)$-code exists.

Proof 'only if' part: Suppose C is a binary (n, M, d)-code,
where d is odd. Let $\hat{C}$ be the code of length $n + 1$ obtained from
C by extending each codeword $\mathbf{x}$ of C according to the rule

$$\mathbf{x} = x_1 x_2 \cdots x_n \rightarrow \hat{\mathbf{x}} = \begin{cases} x_1 x_2 \cdots x_n 0 & \text{if } w(\mathbf{x}) \text{ is even} \\ x_1 x_2 \cdots x_n 1 & \text{if } w(\mathbf{x}) \text{ is odd.} \end{cases}$$

Equivalently we can define

$$\hat{\mathbf{x}} = x_1 x_2 \cdots x_n x_{n+1}$$

where $x_{n+1} = \sum_{i=1}^{n} x_i$, calculated modulo 2.

This construction of $\hat{C}$ from C is called 'adding an overall
parity check' to the code C.

Since $w(\hat{\mathbf{x}})$ is even for every codeword $\hat{\mathbf{x}}$ of $\hat{C}$, it follows from
Lemma 2.6 that $d(\hat{\mathbf{x}}, \hat{\mathbf{y}})$ is even for all $\hat{\mathbf{x}}, \hat{\mathbf{y}}$ in $\hat{C}$. Hence $d(\hat{C})$ is
even. Clearly $d \leq d(\hat{C}) \leq d + 1$, and so, since d is odd, we must
have $d(\hat{C}) = d + 1$. Thus $\hat{C}$ is an $(n + 1, M, d + 1)$-code.

'if' part: Suppose D is an $(n + 1, M, d + 1)$-code, where d is
odd. Choose codewords $\mathbf{x}$ and $\mathbf{y}$ of D such that $d(\mathbf{x}, \mathbf{y}) = d + 1$.
Choose a position in which $\mathbf{x}$ and $\mathbf{y}$ differ and delete this from all
codewords. The result is an (n, M, d)-code.

Corollary 2.8 If d is odd, then $A_2(n + 1, d + 1) = A_2(n, d)$.
Equivalently, if d is even, then $A_2(n, d) = A_2(n - 1, d - 1)$.

Example 2.9 By Example 2.2, $A_2(5, 3) = 4$. Hence, by
Corollary 2.8, $A_2(6, 4) = 4$. To illustrate the 'only if' part of
Theorem 2.7 we construct below a $(6, 4, 4)$-code from the
$(5, 4, 3)$-code of Example 1.5.

$(5, 4, 3)$-code		$(6, 4, 4)$-code
0 0 0 0 0		0 0 0 0 0 0
0 1 1 0 1	$\xrightarrow[\text{parity check}]{\text{add overall}}$	0 1 1 0 1 1
1 0 1 1 0		1 0 1 1 0 1
1 1 0 1 1		1 1 0 1 1 0

The trial-and-error method of Example 2.2, which proved that a binary $(5, M, 3)$-code must have $M \leq 4$, would not be practical for sets of larger parameters. However, there are some general upper bounds on how large a code can be (for given n and d), which sometimes turn out to be the actual value of $A_q(n, d)$. The best known is the so-called 'sphere-packing bound', which we will prove after introducing a little more notation.

Binomial coefficients

If n and m are integers with $0 \leq m \leq n$, then the *binomial coefficient* $\binom{n}{m}$, pronounced 'n choose m', is defined by

$$\binom{n}{m} = \frac{n!}{m! \, (n-m)!},$$

where $m! = m(m-1) \cdots 3.2.1$ for $m > 0$
and $\quad 0! = 1$.

Lemma 2.10 The number of unordered selections of m distinct objects from a set of n distinct objects is $\binom{n}{m}$.

Proof An *ordered* selection of m distinct objects from a set of n distinct objects can be made in

$$n(n-1) \cdots (n-m+1) = \frac{n!}{(n-m)!}$$

ways, for the first object can be chosen in any of n ways, then the second in any of $n-1$ ways, and so on. Since there are $m(m-1) \cdots 2.1 = m!$ ways of ordering the m objects chosen, the number of *unordered* selections is

$$\frac{n!}{m! \, (n-m)!}.$$

Examples 2.11 (i) We illustrate the proof of Lemma 2.10 by listing the ordered and unordered selections of 2 objects from 4. Labelling the four objects 1, 2, 3, 4, the ordered selections of 2 from 4 are $(1, 2)$, $(1, 3)$, $(1, 4)$, $(2, 1)$, $(2, 3)$, $(2, 4)$, $(3, 1)$,

$(3, 2), (3, 4), (4, 1), (4, 2), (4, 3)$. The number of them is $12 = 4.3 = 4!/2!$.

The unordered selections of 2 from 4 are $\{1, 2\}$, $\{1, 3\}$, $\{1, 4\}$, $\{2, 3\}$, $\{2, 4\}$, $\{3, 4\}$. Each unordered selection corresponds to $2! = 2$ ordered selections and so the number of unordered selections is $\dfrac{4!}{2!\,2!} = \binom{4}{2} = 6$.

Note that the unordered selections of m objects from a set S are just the subsets of S of order m.

(ii) Suppose a bet on a football pool is to be a selection (unordered) of 8 matches from a large number. The 8 matches are forecast to be draws (ties). A common plan is to select 10 matches and to 'choose any 8 from 10'. The number of bets required is $\binom{10}{8} = 45$.

(iii) The number of different binary codes with $M = 5$ and $n = 5$ is $\binom{32}{5} = 201\,376$. Of course the number of inequivalent codes will be very much smaller than this.

(iv) The number of binary vectors in $(F_2)^n$ of weight i is $\binom{n}{i}$, this being the number of ways of choosing i positions out of n to have 1s. For example, the vectors in $(F_2)^4$ of weight 2 are 1100, 1010, 1001, 0110, 0101, 0011. The one-to-one correspondence with the list of unordered selections in (i) above should be evident.

We now introduce the notion of a sphere in the set $(F_q)^n$. Provided the analogy is not stretched too far, it can be useful to think of $(F_q)^n$ as a space not unlike the three-dimensional real space which we inhabit. The distance between two points of $(F_q)^n$ is of course taken to be the Hamming distance and then the following definition is quite natural.

Definition. For any vector $\mathbf{u}$ in $(F_q)^n$ and any integer $r \geq 0$, the *sphere* of radius r and centre $\mathbf{u}$, denoted $S(\mathbf{u}, r)$, is the set $\{\mathbf{v} \in (F_q)^n \mid d(\mathbf{u}, \mathbf{v}) \leq r\}$.

Remark 2.12 Let us interpret Theorem 1.9(ii) visually. If

$d(C) \geq 2t + 1$, then the spheres of radius t centred on the codewords of C are disjoint (i.e. they have no overlap). For if a vector $\mathbf{y}$ were in both $S(\mathbf{x}, t)$ and $S(\mathbf{x}', t)$, for codewords $\mathbf{x}$ and $\mathbf{x}'$ (see Fig. 2.13), then by the triangle inequality we would have

$$d(\mathbf{x}, \mathbf{x}') \leq d(\mathbf{x}, \mathbf{y}) + d(\mathbf{x}', \mathbf{y}) \leq t + t = 2t,$$

a contradiction to $d(C) \geq 2t + 1$.

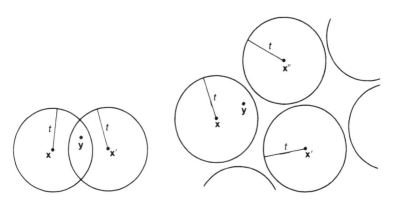

Figure 2.13　　　　　　　**Figure 2.14**

So if t or fewer errors occur in a codeword $\mathbf{x}$, then the received vector $\mathbf{y}$ may be different from the centre of the sphere $S(\mathbf{x}, t)$, but cannot 'escape' from the sphere, and so is 'drawn back' to $\mathbf{x}$ by nearest neighbour decoding (see Fig. 2.14).

Lemma 2.15 A sphere of radius r in $(F_q)^n$ $(0 \leq r \leq n)$ contains exactly

$$\binom{n}{0} + \binom{n}{1}(q-1) + \binom{n}{2}(q-1)^2 + \cdots + \binom{n}{r}(q-1)^r$$

vectors.

Proof Let $\mathbf{u}$ be a fixed vector in $(F_q)^n$. Consider how many vectors $\mathbf{v}$ have distance exactly m from $\mathbf{u}$, where $m \leq n$. The m positions in which $\mathbf{v}$ is to differ from $\mathbf{u}$ can be chosen in $\binom{n}{m}$ ways and then in each of these m positions the entry of $\mathbf{v}$ can be chosen in $q - 1$ ways to differ from the corresponding entry of $\mathbf{u}$.

Hence the number of vectors at distance exactly m from $\mathbf{u}$ is $\binom{n}{m}(q-1)^m$ and so the total number of vectors in $S(\mathbf{u}, r)$ is

$$\binom{n}{0} + \binom{n}{1}(q-1) + \cdots + \binom{n}{r}(q-1)^r.$$

Remark The numbers $\binom{n}{m}$ are called binomial coefficients because of their role in the binomial theorem, which for any positive integer n states that

$$(1+x)^n = 1 + \binom{n}{1}x + \binom{n}{2}x^2 + \cdots + \binom{n}{n}x^n.$$

For x an integer, the binomial theorem follows from Lemma 2.15 by taking $x = q-1$ and $r = n$, for $S(\mathbf{u}, n)$ is the whole of $(F_q)^n$ and so contains $q^n = (1+x)^n$ vectors.

Theorem 2.16 (The *sphere-packing* or *Hamming bound*) A q-ary $(n, M, 2t + 1)$-code satisfies

$$M\left\{\binom{n}{0} + \binom{n}{1}(q-1) + \cdots + \binom{n}{t}(q-1)^t\right\} \leq q^n. \qquad (2.17)$$

Proof Suppose C is a q-ary $(n, M, 2t + 1)$-code. As we observed in Remark 2.12, any two spheres of radius t centred on distinct codewords can have no vectors in common. Hence the total number of vectors in the M spheres of radius t centred on the M codewords is given by the left-hand side of (2.17). This number must be less than or equal to q^n, the total number of vectors in $(F_q)^n$.

For future reference, we re-state (2.17) for the particular case of binary codes. That is, any binary $(n, M, 2t + 1)$-code satisfies

$$M\left\{1 + \binom{n}{1} + \binom{n}{2} + \cdots + \binom{n}{t}\right\} \leq 2^n. \qquad (2.18)$$

For given values of q, n and d, the sphere-packing bound provides an upper bound on $A_q(n, d)$. For example, a binary $(5, M, 3)$-code satisfies $M\{1 + 5\} \leq 2^5 = 32$, and so $A_2(5, 3) \leq 5$. Of course, just because a set of numbers n, M, d satisfies the sphere-packing bound, it does not necessarily mean that a code with those parameters exists. Indeed we saw in Example 2.2 that there is no binary $(5, 5, 3)$-code and that the actual value of $A_2(5, 3)$ is just 4.

Perfect codes

A code which achieves the sphere-packing bound, i.e. such that equality occurs in (2.17), is called a *perfect code*. Thus, for a perfect t-error-correcting code, the M spheres of radius t centred on codewords 'fill' the whole space $(F_q)^n$ without overlapping. Or, in other words, every vector in $(F_q)^n$ is at distance $\leq t$ from exactly one codeword.

The binary repetition code

$$\begin{cases} 0 \ 0 \ \cdots \ 0 \\ 1 \ 1 \ \cdots \ 1 \end{cases}$$

of length n, where n is odd, is a perfect $(n, 2, n)$-code. Such codes, together with codes which contain just one codeword or which are the whole of $(F_q)^n$, are known as *trivial* perfect codes.

The problem of finding all perfect codes has provided mathematicians with one of the greatest challenges in coding theory and we shall return to this problem in Chapter 9. We will conclude this chapter by giving, in Example 2.23, an example of a non-trivial perfect code. An alternative construction, as one of the family of so-called perfect Hamming codes, will be given in Chapter 8, while the present construction will be generalized in Exercise 2.15 to a class of binary codes known as Hadamard codes.

The construction given here will be based on one of a family of configurations known as block designs, which we now introduce.

Balanced block designs

Definition A *balanced block design* consists of a set S of v elements, called *points* or *varieties*, and a collection of b subsets of S, called *blocks*, such that, for some fixed k, r and λ
(1) each block contains exactly k points
(2) each point lies in exactly r blocks
(3) each pair of points occurs together in exactly λ blocks.
Such a design is referred to as a (b, v, r, k, λ)-*design*.

Example 2.19 Take $S = \{1, 2, 3, 4, 5, 6, 7\}$ and consider the following subsets of S: $\{1, 2, 4\}$, $\{2, 3, 5\}$, $\{3, 4, 6\}$, $\{4, 5, 7\}$, $\{5, 6, 1\}$, $\{6, 7, 2\}$, $\{7, 1, 3\}$.

It is easily verified that each pair of elements of S occurs together in exactly one block. Thus the subsets form the blocks of a $(7, 7, 3, 3, 1)$-design.

There is a simple geometrical representation of this design (see Fig. 2.20). The elements $1, 2, \ldots, 7$ are represented by points and the blocks by lines (6 straight lines and a circle). This is known as the *seven-point plane*, the *Fano plane*, or the *projective plane of order 2*.

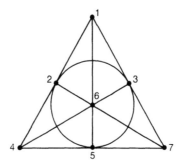

Fig. 2.20 The seven-point plane

The elements of the set S of a block design are often called *varieties* because such designs were originally used in statistical experiments, particularly in agriculture. For example, suppose that we have v varieties of fertilizer to be tested on b crops and that we are particularly interested in the effects of pairs of fertilizers acting together on the same crop. By using a balanced block design, each of the b crops can be tested with a block of k varieties of fertilizer, in such a way that each pair of varieties is tested together a constant number λ of times. Thus the design is balanced so far as comparison between pairs of fertilizers is concerned.

Example 2.21 If we have 7 varieties of fertilizer (labelled $1, 2, \ldots, 7$) and 7 crops, then, using the $(7, 7, 3, 3, 1)$-design of Example 2.19, we could treat the first crop with the block of varieties $\{1, 2, 4\}$, the second crop with $\{2, 3, 5\}$ and so on. The schedule can be displayed as follows:

Figure 2.22

		Blocks						
		B_1	B_2	B_3	B_4	B_5	B_6	B_7
	1	1	0	0	0	1	0	1
	2	1	1	0	0	0	1	0
	3	0	1	1	0	0	0	1
Varieties	4	1	0	1	1	0	0	0
	5	0	1	0	1	1	0	0
	6	0	0	1	0	1	1	0
	7	0	0	0	1	0	1	1

The 7×7 matrix of 0s and 1s thus obtained is called an *incidence matrix* of the design. More formally we have:

Definition The *incidence matrix* $A = [a_{ij}]$ of a block design is a $v \times b$ matrix in which the rows correspond to the varieties $x_1, x_2, \ldots, x_v$ and the columns to the blocks $B_1, B_2, \ldots, B_b$, and whose i, jth entry is defined by

$$a_{ij} = \begin{cases} 1 & \text{if } x_i \in B_j \\ 0 & \text{if } x_i \notin B_j \end{cases}.$$

We now construct our example of a non-trivial perfect code.

Example 2.23 Let A be the incidence matrix of Fig. 2.22 and let B be the 7×7 matrix obtained from A by replacing all 0s by 1s and all 1s by 0s. Let C be the length 7 code whose 16 codewords are the rows $a_1, a_2, \ldots, a_7$ of A, the rows $b_1, b_2, \ldots, b_7$ of B and the additional vectors $\mathbf{0} = 0000000$ and $\mathbf{1} = 1111111$. Thus

$C = 0\ 0\ 0\ 0\ 0\ 0\ 0 = \mathbf{0}$ $0\ 1\ 0\ 1\ 1\ 0\ 0 = a_5$ $1\ 0\ 0\ 1\ 1\ 1\ 0 = b_3$
     $1\ 1\ 1\ 1\ 1\ 1\ 1 = \mathbf{1}$ $0\ 0\ 1\ 0\ 1\ 1\ 0 = a_6$ $0\ 1\ 0\ 0\ 1\ 1\ 1 = b_4$
     $1\ 0\ 0\ 0\ 1\ 0\ 1 = a_1$ $0\ 0\ 0\ 1\ 0\ 1\ 1 = a_7$ $1\ 0\ 1\ 0\ 0\ 1\ 1 = b_5$
     $1\ 1\ 0\ 0\ 0\ 1\ 0 = a_2$ $0\ 1\ 1\ 1\ 0\ 1\ 0 = b_1$ $1\ 1\ 0\ 1\ 0\ 0\ 1 = b_6$
     $0\ 1\ 1\ 0\ 0\ 0\ 1 = a_3$ $0\ 0\ 1\ 1\ 1\ 0\ 1 = b_2$ $1\ 1\ 1\ 0\ 1\ 0\ 0 = b_7$
     $1\ 0\ 1\ 1\ 0\ 0\ 0 = a_4$

We will show that the minimum distance of C is 3, i.e. that $d(\mathbf{x}, \mathbf{y}) \geq 3$ for any pair of codewords $\mathbf{x}, \mathbf{y}$. By the incidence properties of the $(7, 7, 3, 3, 1)$-design, each row of A has exactly 3 1s and any two distinct rows of A have exactly one 1 in common. Hence, by Lemma 2.6,

$$d(\mathbf{a}_i, \mathbf{a}_j) = 3 + 3 - 2.1 = 4 \qquad \text{for } i \neq j.$$

Since distances between codewords are unchanged if all 0s are changed to 1s and all 1s to 0s, we have also that

$$d(\mathbf{b}_i, \mathbf{b}_j) = 4 \qquad \text{for } i \neq j.$$

It is clear that

$$d(\mathbf{0}, \mathbf{y}) \;= 3, 4 \text{ or } 7 \text{ according as } \mathbf{y} = \mathbf{a}_i, \mathbf{b}_j \text{ or } \mathbf{1},$$
$$d(\mathbf{1}, \mathbf{y}) \;= 3, 4 \text{ or } 7 \text{ according as } \mathbf{y} = \mathbf{b}_i, \mathbf{a}_j \text{ or } \mathbf{0},$$
$$\text{and} \qquad d(\mathbf{a}_i, \mathbf{b}_i) = 7 \qquad \text{for } i = 1, 2, \ldots, 7.$$

It remains only to consider $d(\mathbf{a}_i, \mathbf{b}_j)$ for $i \neq j$. But $\mathbf{a}_i$ and $\mathbf{b}_j$ differ precisely in those places where $\mathbf{a}_i$ and $\mathbf{a}_j$ agree and so

$$d(\mathbf{a}_i, \mathbf{b}_j) = 7 - d(\mathbf{a}_i, \mathbf{a}_j) = 7 - 4 = 3.$$

We have now shown that C is a $(7, 16, 3)$-code and since

$$16\left(\binom{7}{0} + \binom{7}{1}\right) = 2^7,$$

we have equality in (2.18) and so the code is perfect.

The existence of a perfect binary $(7, 16, 3)$-code shows that $A_2(7, 3) = 16$ and so we have established another of the entries of Table 2.4.

In leaving the code of Example 2.23 we note that it has the remarkable property that the sum of any two codewords is also a codeword! Interestingly, the $(5, 4, 3)$-code of Example 2.2 has the same property. Such codes are called *linear codes* and play a central role in coding theory. We shall begin to study the theory of such codes in Chapter 5.

Concluding remarks on Chapter 2

(1) It is not recommended that the reader spends a lot of time on the unresolved cases in Table 2.4, for many man-hours have so far failed to improve on the current best bounds.

However, the manner in which one entry, $A_2(15, 5) = 256$, was obtained (Nordstrom and Robinson 1967) might give some encouragement to the amateur. It was previously known only that $128 \leq A_2(15, 5) \leq 256$ and this case was chosen by Robinson as an example of a problem which he posed to high school students in an introductory talk on coding theory. One of them, named Alan Nordstrom, accepted the challenge and, by trial and error, constructed a $(15, 256, 5)$-code, the now-famous Nordstrom–Robinson code. A construction of this code will be given in Exercise 9.9.

It might be felt that all optimal codes of moderate length should be obtainable by means of exhaustive computer searches. But an estimate of the time needed to find whether there exists, say, a binary $(10, 73, 3)$-code shows how difficult this would be. In fact, computer-aided searches have so far met with distinctly limited success; almost all the good codes known have arisen out of their discoverers' ingenuity.

(2) For binary codes, the sphere-packing bound turns out to be reasonably good for cases $n \geq 2d + 1$. Unfortunately, it becomes very weak for $n < 2d$, but in such cases there is a much sharper bound, due to Plotkin (1960), which will be derived in Exercises 2.20–22. [For some recent analogous results on ternary codes, see Mackenzie and Seberry (1984). For some bounds on binary (n, M, d)-codes with n slightly greater than $2d$, see Tietäväinen (1980).]

The reader who wishes to progress quickly to the main stream of coding theory, which is the theory of linear codes, need not dwell on the remaining remarks of this chapter for too long and may also leave Exercises 2.12 to 2.24 for the time being.

(3) The parameters of a (b, v, r, k, λ)-design are not independent, for they satisfy the following two conditions (see Exercise 2.13):

$$bk = vr \qquad (2.24)$$

$$r(k - 1) = \lambda(v - 1). \qquad (2.25)$$

However, if five numbers b, v, r, k, λ satisfy (2.24) and (2.25), there is no guarantee that a (b, v, r, k, λ)-design exists. For example it is known that there does not exist a $(43, 43, 7, 7, 1)$-design.

(4) A block design is called *symmetric* if $v = b$ (and so also, by (2.24), $k = r$), and is referred to simply as a (v, k, λ)-*design*. There are two types of (v, k, λ)-design which will be of particular interest to us.

(i) A *finite projective plane* is a symmetric design for which $\lambda = 1$. If we put $k = n + 1$, then n is called the *order* of the plane. By (2.25), we then have $v = n^2 + n + 1$, and so a *projective plane of order* n is a $(n^2 + n + 1, n + 1, 1)$-design. Such a design exists whenever n is a prime power (see Exercise 4.7).

(ii) A $(4t - 1, 2t - 1, t - 1)$-design is called a *Hadamard design*.

We see that the $(7, 3, 1)$-design of Example 2.19 is both a projective-plane of order 2 and a Hadamard design with $t = 2$.

(5) Further relations on the five parameters of a (b, v, r, k, λ)-design have been found by making ingenious use of the incidence matrix. The best known is the very simple, but by no means obvious, result that

$$v \leq b \qquad (2.26)$$

obtained by the statistician R. A. Fisher in 1940.

For the particular case of symmetric designs, the following fundamental theorem was proved by Bruck, Ryser and Chowla in 1950.

Theorem 2.27 If a (v, k, λ)-design exists, then
(i)   if v is even, $k - \lambda$ is a square
(ii) if v is odd, the equation $z^2 = (k - \lambda)x^2 + (-1)^{(v-1)/2}\lambda y^2$ has a solution in integers x, y, z not all zero.

It is an unsolved problem to determine whether the necessary condition of Theorem 2.27, together with (2.24) and (2.25), form a set of *sufficient* conditions for the existence of a symmetric design. There are many parameters for which the existence of the design is undecided, a particularly interesting case being the projective plane of order 10, with parameters $(v, k, \lambda) = (111, 11, 1)$.

For full details of these, and other, results on block designs the reader is referred to Anderson (1974) or Hall (1980).

(6) A generalization of block designs to so-called '*t*-designs' will be considered in Chapter 9.

Exercises 2

Questions should *not* be answered simply by referring to Table 2.4.

2.1 Construct, if possible, binary (n, M, d)-codes with the following parameters: $(6, 2, 6)$, $(3, 8, 1)$, $(4, 8, 2)$, $(5, 3, 4)$, $(8, 30, 3)$. (When not possible, show *why* not possible).

2.2 Show that if there exists a binary (n, M, d)-code, then there exists a binary $(n - 1, M', d)$-code with $M' \geqslant M/2$. Deduce that $A_2(n, d) \leqslant 2A_2(n - 1, d)$.

2.3 Prove that $A_q(3, 2) = q^2$ for any integer $q \geqslant 2$. [*Hint*: See Exercise 1.5].

2.4 Let E_n denote the set of all vectors in $(F_2)^n$ which have even weight. Show that E_n is the code obtained by adding an overall parity check to the code $(F_2)^{n-1}$. Deduce that E_n is an $(n, 2^{n-1}, 2)$-code.

2.5 Consider an entry to a football pool made by selecting 10 matches at random from a total of 50 and 'choosing any 8 from 10'. Show that if exactly 8 of the 50 matches finish as draws, the odds against the above entry containing a winning line are greater than 10 million to 1.

2.6 Show that if there is a binary (n, M, d)-code with d even, then there exists a binary (n, M, d)-code in which all the codewords have even weight.

2.7 Show that the number of inequivalent binary codes of length n and containing just two codewords is n.

2.8 Show that $A_2(8, 5) = 4$ and that, up to equivalence, there is just one binary $(8, 4, 5)$-code.

2.9 Show that any q-ary (n, q, n)-code is equivalent to a repetition code.

2.10 Show that a q-ary $(q + 1, M, 3)$-code satisfies $M \leqslant q^{q-1}$.

2.11 Show that $A_2(8, 4) = 16$.

2.12 Listed below are the blocks of an $(11, 5, 2)$-design. Use this to construct a binary $(11, 24, 5)$-code.

$\{1, 3, 4, 5, 9\}$, $\{2, 4, 5, 6, 10\}$, $\{3, 5, 6, 7, 11\}$,
$\{1, 4, 6, 7, 8\}$, $\{2, 5, 7, 8, 9\}$, $\{3, 6, 8, 9, 10\}$,
$\{4, 7, 9, 10, 11\}$, $\{1, 5, 8, 10, 11\}$, $\{1, 2, 6, 9, 11\}$,
$\{1, 2, 3, 7, 10\}$, $\{2, 3, 4, 8, 11\}$.

[*Remark*: We see from Table 2.4 that $A_2(11, 5) = 24$ and

so the code constructed here is the largest binary double-error-correcting code of length 11. We shall prove this in Exercise 2.22(iv).]

Show that the sphere-packing bound for a binary $(11, M, 5)$-code gives only $M \le 30$.

2.13 Show that the parameters of a (b, v, r, k, λ)-design satisfy (i) $bk = vr$, (ii) $r(k - 1) = \lambda(v - 1)$. [*Hint* for (i): Count in two ways the number of ordered pairs in the set $\{(x, B): x$ is a point, B is a block and $x \in B\}$.]

2.14 Show that there do not exist (b, v, r, k, λ)-designs with the parameters: (i) $(12, 8, 6, 4, 3)$, (ii) $(22, 22, 7, 7, 2)$.

2.15 Show that if there exists a Hadamard $(4t - 1, 2t - 1, t - 1)$-design, then $A_2(4t - 1, 2t - 1) \ge 8t$.

2.16 Let C be the binary code consisting of all cyclic shifts of the vectors 11010000, 11100100 and 10101010, together with $\mathbf{0}$ and $\mathbf{1}$. (A cyclic shift of $a_1 a_2 \cdots a_n$ is a vector of the form $a_i a_{i+1} \cdots a_n a_1 a_2 \cdots a_{i-1}$.) Show that C is a $(8, 20, 3)$-code. When showing that $d(C) = 3$, the cyclic nature of the code reduces the number of evaluations of $d(\mathbf{x}, \mathbf{y})$ required from $\binom{20}{2}$ to $\cdots$ (how many?).

2.17 [The $(\mathbf{u} \mid \mathbf{u} + \mathbf{v})$ construction of Plotkin (1960).] Given $\mathbf{u} = u_1 \cdots u_m$ and $\mathbf{v} = v_1 \cdots v_n$, let $(\mathbf{u} \mid \mathbf{v})$ denote the vector $u_1 \cdots u_m v_1 \cdots v_n$ of length $m + n$. Suppose that C_1 is a binary (n, M_1, d_1)-code and that C_2 is a binary (n, M_2, d_2)-code. Form a new code C_3 consisting of all vectors of the form $(\mathbf{u} \mid \mathbf{u} + \mathbf{v})$, where $\mathbf{u} \in C_1$, $\mathbf{v} \in C_2$. Show that C_3 is a $(2n, M_1 M_2, d)$-code with $d = \min \{2d_1, d_2\}$.

2.18 Prove that $A_2(16, 3) \ge 2560$. [*Hint:* Use Exercises 2.16 and 2.17.]

2.19 Starting from the $(4, 8, 2)$ even-weight code (see Exercise 2.4) and the $(4, 2, 4)$ repetition code, apply Exercise 2.17 three times to show that there exists a binary $(32, 64, 16)$-code. [*Remark:* The $(2^m, 2^{m+1}, 2^{m-1})$-codes, which may be constructed in this way for each positive integer $m \ge 1$, are called first-order *Reed–Muller* codes.]

The aim of the next three exercises is to derive the so-called Plotkin bound.

2.20 Show that if C is a binary (n, M, d)-code with $n < 2d$, then

$$M \leq \begin{cases} 2d/(2d - n) & \text{if } M \text{ is even} \\ 2d/(2d - n) - 1 & \text{if } M \text{ is odd.} \end{cases}$$

[*Hint:* let $C = \{\mathbf{x}_1, \mathbf{x}_2, \ldots, \mathbf{x}_M\}$ and let T be the $\binom{M}{2} \times n$ matrix whose rows are the vectors $\mathbf{x}_i + \mathbf{x}_j$, $1 \leq i < j \leq M$. Estimate the number $w(T)$ of non-zero entries of T in two ways, via rows and via columns.]

2.21 Deduce from Exercise 2.20 that, if $n < 2d$, then

$$A_2(n, d) \leq 2\lfloor d/(2d - n) \rfloor.$$

State the upper bounds this gives on $A_2(9, 5)$ and on $A_2(10, 6)$. How can the bound on $A_2(9, 5)$ be improved? [*Remark:* As for this case, it happens in general that the above bound is good for d even, but is open to improvement for d odd; we make that improvement in the next exercise.]

2.22 Show that
   (i) if d is even and $n < 2d$, then

$$A_2(n, d) \leq 2\lfloor d/(2d - n) \rfloor,$$

   (ii) if d is odd and $n < 2d + 1$, then

$$A_2(n, d) \leq 2\lfloor (d + 1)/(2d + 1 - n) \rfloor,$$

   (iii) if d is even, then $A_2(2d, d) \leq 4d$,
   (iv) if d is odd, then $A_2(2d + 1, d) \leq 4d + 4$.
   (i) to (iv) are known collectively as the *Plotkin bound*.

2.23 Show that the $(32, 64, 16)$-code of Exercise 2.19 is optimal. Generalize this result by proving that $A_2(2d, d) = 4d$ whenever d is a power of 2.

2.24 Show that if there exists a Hadamard $(4t - 1, 2t - 1, t - 1)$-design, then $A_2(4t, 2t) = 8t$.

3 An introduction to finite fields

To make error-correcting codes easier to use and analyse, it is necessary to impose some algebraic structure on them. It is especially useful to have an alphabet in which it is possible to add, subtract, multiply and divide without restriction. In other words we wish to give F_q the structure of a *field*, the formal definition of which follows.

Definition A *field* F is a set of elements with two operations $+$ (called addition) and $\cdot$ (multiplication) satisfying the following properties.

    (i) F is closed under $+$ and $\cdot$, i.e. $a + b$ and $a \cdot b$ are in F whenever a and b are in F.

For all a, b and c in F, the following laws hold.

    (ii)   Commutative laws: $a + b = b + a$, $a \cdot b = b \cdot a$.

    (iii)   Associative laws: $(a + b) + c = a + (b + c)$, $a \cdot (b \cdot c) = (a \cdot b) \cdot c$.

    (iv)   Distributive law: $a \cdot (b + c) = a \cdot b + a \cdot c$.

Furthermore, identity elements 0 and 1 must exist in F satisfying

    (v) $a + 0 = a$ for all a in F.

    (vi) $a \cdot 1 = a$ for all a in F.

    (vii)   For any a in F, there exists an additive inverse element $(-a)$ in F such that $a + (-a) = 0$.

    (viii)   For any $a \neq 0$ in F, there exists a multiplicative inverse element a^{-1} in F such that $a \cdot a^{-1} = 1$.

Notes

(1) From now on we will generally write $a \cdot b$ simply as ab.

(2) We can regard a field F as having the four operations $+$, $-$, $\cdot$ and $\div$, where $-$ and $\div$ are given by (vii) and (viii) respectively with the understanding that $a - b = a + (-b)$ and $a \div b$, or a/b, $= a(b^{-1})$ for $b \neq 0$.

(3) The reader who has done any group theory will recognize
     that a field can be more concisely defined to be a set of
     elements such that
     (a) it is an abelian group under +,
     (b) the non-zero elements form an abelian group under ·,
     (c) the distributive law holds.
(4) The following two further properties of a field are easily
     deduced from the definition.

Lemma 3.1 Any field F has the following properties.
(i) $a0 = 0$ for all a in F.
(ii) $ab = 0 \Rightarrow a = 0$ or $b = 0$. (Thus the product of two non-zero
      elements of a field is also non-zero.)

Proof (i) We have $a0 = a(0 + 0) = a0 + a0$. Adding the additive inverse of $a0$ to both sides gives

$$0 = a0 + (-a0) = a0 + a0 + (-a0) = a0 + 0 = a0.$$

Thus $a0 = 0$.
   (ii) Suppose $ab = 0$. If $a \neq 0$, then a has a multiplicative
inverse and so $b = 1 \cdot b = (a^{-1}a)b = a^{-1}(ab) = a^{-1}0 = 0$. Hence
$ab = 0 \Rightarrow a = 0$ or $b = 0$.

Definition A set of elements with + and · satisfying the field
properties (i) to (vii), but not necessarily (viii), is called a *ring*.

Remark For convenience, we have defined a 'ring' to be a
structure which should properly be called a 'commutative (or
abelian) ring, with an identity'.
   Familiar examples of infinite fields are the set of real numbers
and the set of complex numbers. The set Z of integers is a ring
but is not a field because, for example, 2 does not have a
multiplicative inverse in Z. Another example of a ring which is
not a field is the set $F[x]$ of polynomials in x with coefficients
belonging to a field F. This ring will be of importance in Chapter
12.

Definition A *finite field* is a field which has a finite number of
elements, this number being called the *order* of the field.
   The following fundamental result about finite fields was proved

by Evariste Galois (1811–32), a French mathematician who died in a duel at the age of 20. Galois is famous also for proving that the general quintic equation is not solvable by radicals.

Theorem 3.2 There exists a field of order q if and only if q is a prime power (i.e. $q = p^h$, where p is prime and h is a positive integer). Furthermore, if q is a prime power, then there is, up to relabelling, only one field of that order.

A field of order q is often called a *Galois field* of order q and is denoted $GF(q)$.

The proof of Theorem 3.2 may be found in one of the more advanced texts on coding theory or in books on abstract algebra. While we shall give a partial proof in Exercise 4.6, and shall give a brief description of fields of order p^h, with $h > 1$, in Chapter 12, it is enough for almost all our purposes to consider only *prime fields*, those of order a prime number p. We shall see shortly that if p is prime, then $GF(p)$ is just the set $\{0, 1, \ldots, p - 1\}$ with arithmetic carried out modulo p. But first we review modular arithmetic in general.

Definition Let m be a fixed positive integer. Two integers a and b are said to be *congruent* (*modulo m*), symbolized by

$$a \equiv b \pmod{m},$$

if $a - b$ is divisible by m, i.e. if $a = km + b$ for some integer k.

We write $a \not\equiv b \pmod{m}$ if a and b are not congruent (modulo m).

Every integer, when divided by m, has a unique principal remainder equal to one of the integers in the set $Z_m = \{0, 1, \ldots, m - 1\}$. It is easily shown that two integers are congruent $(\bmod\, m)$ if and only if they have the same principal remainders on division by m.

Examples $3 \equiv 24 \pmod{7}$, $13 \equiv -2 \pmod{5}$, $25 \not\equiv 12 \pmod{7}$,
$15 \equiv 0 \pmod{3}$, $15 \equiv 0 \pmod{5}$, $15 \not\equiv 0 \pmod{2}$.

Theorem 3.3 Suppose $a \equiv a' \pmod{m}$ and $b \equiv b' \pmod{m}$. Then
(i) $a + b \equiv a' + b' \pmod{m}$
(ii) $ab \equiv a'b' \pmod{m}$.

Proof $a = a' + km$ and $b = b' + lm$ for some integers k and l. Then (i) $a + b = a' + b' + (k + l)m$ and so $a + b \equiv a' + b' \pmod{m}$ and (ii) $ab = a'b' + (kb' + a'l + klm)m$ and so $ab \equiv a'b' \pmod{m}$.

Theorem 3.3 enables congruences to be calculated without working with large numbers. Note that if $a \equiv a'$, then repeated use of (ii) shows that, for all positive integers n, $a^n \equiv (a')^n \pmod{m}$.

Examples 3.4 (i) What is the principal remainder when $73 \cdot 52$ is divided by 7?

(ii) Determine whether $(2^{15})(14^{40}) + 1$ is divisible by 11.

Solution (i) $73 \equiv 3 \pmod{7}$ and $52 \equiv 3 \pmod{7}$. Hence, by Theorem 3.3(ii), $73 \cdot 52 \equiv 3 \cdot 3 \equiv 9 \equiv 2 \pmod{7}$. So the principal remainder is 2. (There is no need actually to multiply 73 by 52 and divide the answer by 7.)

(ii) Note that $2^5 \equiv 32 \equiv -1 \pmod{11}$. Also $14^2 \equiv 3^2 \equiv -2 \pmod{11}$. Hence

$$(2^{15})(14^{40}) \equiv (2^5)^3(3^2)^{20} \equiv (-1)^3(-2)^{20}$$
$$\equiv (-1)(2^{20}) \equiv (-1)(2^5)^4 \equiv (-1)(-1)^4 \equiv -1 \pmod{11}.$$

Thus $(2^{15})(14^{40}) + 1 \equiv 0 \pmod{11}$, i.e. the number is divisible by 11.

Let us now try to give $Z_m = \{0, 1, \dots, m - 1\}$ the structure of a field. We define *addition* and *multiplication* in Z_m by: $a + b$ (or ab) = the principal remainder when $a + b$ (or ab) is divided by m.

For example, in Z_{12} we have

$$8 + 4 = 0, \quad 9 + 11 = 8, \quad 3 \cdot 4 = 0, \quad 3 \cdot 9 = 3.$$

Theorem 3.3 shows that addition and multiplication in Z_m are well-defined and it is easily verified that the field properties (i) to (vii) are satisfied for any m (the additive inverse of a is $m - a$). Thus, for any integer $m \geqslant 2$, Z_m is a ring. It is called the *ring of integers modulo m*. But for which values of m is field property (viii) satisfied? The following theorem gives the answer.

Theorem 3.5 Z_m is a field if and only if m is a prime number.

Proof First, suppose m is not prime. Then $m = ab$ for some integers a and b, both less than m. Thus

$$ab \equiv 0 \pmod{m}, \quad \text{with } a \not\equiv 0 \pmod{m} \text{ and } b \not\equiv 0 \pmod{m}.$$

So, in Z_m, the product of the non-zero elements a and b is zero and so, by Lemma 3.1(ii), Z_m is not a field.

Now suppose that m is prime. By the remarks preceding this theorem, to show that Z_m is a field it is enough to show that every non-zero element of Z_m has a multiplicative inverse. Let a be a non-zero element of Z_m and consider the $m - 1$ elements $1a, 2a, \ldots, (m-1)a$. These elements are non-zero, for ia cannot have the prime m as a divisor if i and a do not. Also the elements are distinct from one another, for

$$ia = ja \Rightarrow (i - j)a \equiv 0 \pmod{m}$$

$$\Rightarrow m \text{ is a divisor of } (i - j)a$$

$$\Rightarrow m \text{ is a divisor of } i - j, \text{ since } m \text{ is prime}$$
and does not divide a.

$$\Rightarrow i = j, \text{ since both } i \text{ and } j \in \{1, 2, \ldots, m-1\}.$$

So, in Z_m, the $m - 1$ elements $1a, 2a, \ldots, (m-1)a$ must be equal to the elements $1, 2, \ldots, m-1$, in some order, and one of them, ja say, must be equal to 1. This j is the desired inverse of a.

Examples 3.6 (1) $GF(2) = Z_2 = \{0, 1\}$ with addition and multiplication tables

+	0 1
0	0 1
1	1 0

·	0 1
0	0 0
1	0 1

(2) $GF(3) = Z_3 = \{0, 1, 2\}$ with tables

+	0 1 2
0	0 1 2
1	1 2 0
2	2 0 1

·	0 1 2
0	0 0 0
1	0 1 2
2	0 2 1

(3) Z_4 is not a field by Theorem 3.5 (examination of the multiplication table of Z_4 shows that 2 does not have an inverse and so we cannot divide by 2 in Z_4). However, while $4 = 2^2$ is not prime, it is a prime power, and so the field $GF(4)$ does exist, by Theorem 3.2. It can be defined as $GF(4) = \{0, 1, a, b\}$ with tables

+	0 1 a b
0	0 1 a b
1	1 0 b a
a	a b 0 1
b	b a 1 0

·	0 1 a b
0	0 0 0 0
1	0 1 a b
a	0 a b 1
b	0 b 1 a

We shall meet this field in its natural setting in Example 12.2.

(4) Z_6 and Z_{10} are not fields, nor is there *any* field of order 6 or 10.

(5) $GF(11) = Z_{11} = \{0, 1, 2, \ldots, 10\}$ is a field. We can easily carry out addition, subtraction and multiplication (modulo 11) without using tables. But what about division? Remember, to divide a by b, we just multiply a by b^{-1}. So how do we find b^{-1}? The proof of Theorem 3.5 shows the existence of multiplicative inverses but not how to find them efficiently. Two methods for a general prime modulus m are described in Exercises 3.8 and 3.9. For a modulus as small as $m = 11$ it is easy to construct, by trial and error, a table of inverses, thus:

x	1 2 3 4 5 6 7 8 9 10
x^{-1}	1 6 4 3 9 2 8 7 5 10

To illustrate the use of this table, we will divide 6 by 8 in the field $GF(11)$. We have
$$\tfrac{6}{8} = 6 \cdot 8^{-1} = 6 \cdot 7 = 42 = 9.$$

We can give an immediate application of the use of modulo 11 arithmetic in an error-detecting code.

The ISBN code

Every recent book should have an International Standard Book Number (ISBN). This is a 10-digit codeword assigned by the publisher. For example, a book might have the ISBN

0-19-859617-0

although the hyphens may appear in different places and are in fact unimportant. The first digit, 0, indicates the language (English) and the next two digits 19 stand for Oxford University Press, the publishers. The next six digits 859617 are the book number assigned by the publisher, and the final digit is chosen to make the whole 10-digit number $x_1x_2 \cdots x_{10}$ satisfy

$$\sum_{i=1}^{10} ix_i \equiv 0 \pmod{11}. \tag{3.7}$$

The left-hand side of (3.7) is called the *weighted check sum* of the number $x_1x_2 \cdots x_{10}$. Thus for the 9-digit number $x_1x_2 \cdots x_9$ already chosen, x_{10} is defined by

$$x_{10} = \sum_{i=1}^{9} ix_i \pmod{11}$$

to get the ISBN.

The publisher is forced to allow a symbol X in the final position if the check digit x_{10} turns out to be a '10'; e.g. *Chambers Twentieth Century Dictionary* has ISBN 0550-10206-X.

The ISBN code is designed to *detect* (a) any single error and (b) any double-error created by the transposition of two digits. The error detection scheme is simply this. For a received vector $y_1y_2 \cdots y_{10}$ calculate its weighted check sum $Y = \sum_{i=1}^{10} iy_i$. If $Y \neq 0 \pmod{11}$, then we have detected error(s). Let us verify that this works for cases (a) and (b) above. Suppose $\mathbf{x} = x_1x_2 \cdots x_{10}$ is the codeword sent.

(a) Suppose the received vector $\mathbf{y} = y_1y_2 \cdots y_{10}$ is the same as $\mathbf{x}$ except that digit x_j is received as $x_j + a$ with $a \neq 0$. Then $Y = \sum_{i=1}^{10} iy_i = (\sum_{i=1}^{10} ix_i) + ja = ja \neq 0 \pmod{11}$, since j and a are non-zero.

(b) Suppose $\mathbf{y}$ is the same as $\mathbf{x}$ except that digits x_j and x_k have been transposed. Then

$$Y = \sum_{i=1}^{10} iy_i = \sum_{i=1}^{10} ix_i + (k-j)x_j + (j-k)x_k$$
$$= (k-j)(x_j - x_k) \neq 0 \pmod{11},$$
$$\text{if } k \neq j \text{ and } x_j \neq x_k.$$

Note how crucial use is made of the result (Lemma 3.1(ii)) that in a field, the product of two non-zero elements is also non-zero. This does not hold in Z_{10}, in which, for example,

$2 \cdot 5 \equiv 0 \pmod{10}$, and this is why we work with modulus 11 rather than 10. We shall discuss some further codes based on modulo 11 arithmetic in Chapters 7 and 11.

The ISBN code cannot be used to *correct* an error unless we know that just one *given* digit is in error. This is the basis of the following party trick.

Ask a friend to choose a book not known to you and to read out its ISBN, but saying '*x*' for one of the digits. After a few seconds working you announce the value of *x*. For example, if the number read out is 0-201-1*x*-502-7, your working is:

$$1 \cdot 0 + 2 \cdot 2 + 3 \cdot 0 + 4 \cdot 1 + 5 \cdot 1 + 6 \cdot x + 7 \cdot 5 + 8 \cdot 0 + 9 \cdot 2 + 10 \cdot 7 = 0.$$

Hence $6x + 4 = 0$, and so

$$x = \frac{-4}{6} = 7 \cdot 6^{-1} = 7 \cdot 2 = 14 = 3.$$

Concluding Remark   It is hoped that the reader is beginning to appreciate the power and versatility of finite fields, which the author believes to be among the most beautiful structures in mathematics. One remarkable property of any finite field, not needed in this book and so not proved here, is that all the non-zero elements can be expressed as powers of a single element, which is called a *primitive element*; i.e. there exists $g \in GF(q)$ such that the non-zero elements of $GF(q)$ are precisely $1, g, g^2, \ldots, g^{q-2}$, with $g^{q-1} = 1$. This result is by no means obvious, even if we restrict our attention to the case of prime fields. One application of this result is that in a large or complicated field a table of indices of the non-zero elements, with respect to a fixed primitive root, can be constructed, and this can be used, in the same way as logarithms, to carry out multiplication in the field.

For an encyclopaedic volume on finite fields the reader is referred to Lidl and Niederreiter (1983).

Exercises 3

3.1   Find the principal remainder when 2^{20} is divided by 7. Find the units digit of 3^{100}.

3.2   Show that every square integer is congruent $\pmod 4$ to either 0 or 1. Hence show that there do not exist integers x and y such that $x^2 + y^2 = 1839$.

3.3 Construct a table of multiplicative inverses for (i) $GF(7)$, (ii) $GF(13)$.

3.4 (i) What is the minimum distance of the ISBN code?
(ii) What proportion of books would you expect to have an ISBN containing the symbol X?

3.5 Check whether the following are ISBNs.

$$0\text{-}13165332\text{-}6$$
$$0\text{-}1392\text{-}4101\text{-}4$$
$$07\text{-}028761\text{-}4$$

3.6 The following ISBNs have been received with smudges. What are the missing digits?

$$0\text{-}13\text{-}1\blacksquare9139\text{-}9$$
$$0\text{-}02\text{-}32\blacksquare\blacksquare80\text{-}0$$

3.7 Consider the code C of all 10-digit numbers over the 10-ary alphabet $\{0, 1, \ldots, 9\}$ which have the property that the sum of their digits is divisible by 11; that is,

$$C = \left\{ x_1 x_2 \cdots x_{10} \,\middle|\, \sum_{i=1}^{10} x_i \equiv 0 \pmod{11} \right\}.$$

Show that C can detect any single error. What would be the disadvantage of using this code for book numbers rather than the ISBN code?

3.8 Let a be a non-zero element of $GF(p)$, where p is prime. By considering the product of the $p-1$ elements $1a, 2a, \ldots, (p-1)a$, prove that

$$a^{p-1} \equiv 1 \pmod{p} \qquad \text{(Fermat's theorem)}.$$

Deduce that $a^{-1} \equiv a^{p-2} \pmod{p}$. [*Remark:* for p large, a more efficient method of finding a^{-1} is given in the next exercise].

3.9 The Euclidean algorithm is a well-known method of finding the greatest common divisor d of two integers a and b. It also enables d to be expressed in the form

$$d = ax + by$$

for some integers x and y. Show that the Euclidean algorithm can therefore be used to find the inverse of an element $a \neq 0$ in the field $GF(p)$, where p is prime. If you know the Euclidean algorithm, use it to calculate $23^{-1} \pmod{31}$.

3.10 Find a primitive element for each of $GF(3)$, $GF(7)$ and $GF(11)$.

3.11 Suppose F is a finite field. Given that $\alpha \in F$ and n is a positive integer, let $n\alpha$ denote the element $\alpha + \alpha + \cdots + \alpha$ (n terms). Prove that there exists a prime number p such that $p\alpha = 0$ for all $\alpha \in F$. This prime number p is called the *characteristic* of the field F.

3.12 Suppose p is a prime number. Show that $(a + b)^p \equiv a^p + b^p \pmod{p}$. [*Hint:* show that $\binom{p}{i} \equiv 0 \pmod{p}$ if $1 \le i \le p - 1$.] Deduce that $a^p \equiv a \pmod{p}$, for any integer a. (This gives an alternative proof of Fermat's theorem, Exercise 3.8.)

3.13 In the field $GF(q)$, where q is odd, show that the product of all the non-zero elements is equal to -1.

3.14 Show that in a finite field of characteristic p,
   (i) if $p = 2$, then every element is a square
   (ii) if p is odd, then exactly half of the non-zero elements are squares.

4 Vector spaces over finite fields

In addition to carrying out arithmetical operations within the alphabet of a code, it is also very useful to be able to perform certain operations with the codewords themselves. We have already benefited from this in making use of the 'sum' of two binary vectors to prove Lemma 2.6.

Throughout this chapter we assume that q is a prime power and we let $GF(q)$ denote the finite field of q elements. The elements of $GF(q)$ will be called *scalars*. The set $GF(q)^n$ of all ordered n-tuples over $GF(q)$ will now be denoted by $V(n, q)$ and its elements will be called *vectors*.

We define two operations within $V(n, q)$:

(i) *addition of vectors*: if $\mathbf{x} = (x_1, x_2, \ldots, x_n)$ and $\mathbf{y} = (y_1, y_2, \ldots, y_n) \in V(n, q)$, then

$$\mathbf{x} + \mathbf{y} = (x_1 + y_1, x_2 + y_2, \ldots, x_n + y_n)$$

(ii) *multiplication of a vector by a scalar*: if

$$\mathbf{x} = (x_1, x_2, \ldots, x_n) \in V(n, q) \quad \text{and} \quad a \in GF(q),$$

then $a\mathbf{x} = (ax_1, ax_2, \ldots, ax_n)$.

The reader should have no difficulty in verifying that $V(n, q)$ satisfies the axioms for a *vector space*; i.e. that, for all $\mathbf{u}, \mathbf{v}, \mathbf{w} \in V(n, q)$ and for all $a, b \in GF(q)$,

(i) $\mathbf{u} + \mathbf{v} \in V(n, q)$

(ii) $(\mathbf{u} + \mathbf{v}) + \mathbf{w} = \mathbf{u} + (\mathbf{v} + \mathbf{w})$

(iii) the all-zero vector $\mathbf{0} = (0, 0, \ldots, 0) \in V(n, q)$ and satisfies $\mathbf{u} + \mathbf{0} = \mathbf{0} + \mathbf{u} = \mathbf{u}$.

(iv) Given $\mathbf{u} = (u_1, u_2, \ldots, u_n) \in V(n, q)$, the element $-\mathbf{u} = (-u_1, -u_2, \ldots, -u_n) \in V(n, q)$ and satisfies $\mathbf{u} + (-\mathbf{u}) = \mathbf{0}$.

(v) $\mathbf{u} + \mathbf{v} = \mathbf{v} + \mathbf{u}$.
(Properties (i)–(v) mean that $V(n, q)$ is an 'abelian group' under addition).

   (vi) (Closure under scalar multiplication) $a\mathbf{v} \in V(n,q)$.
  (vii) (Distributive laws) $a(\mathbf{u}+\mathbf{v}) = a\mathbf{u}+a\mathbf{v}$, $(a+b)\mathbf{u} = a\mathbf{u}+b\mathbf{u}$.
 (viii) $(ab)\mathbf{u} = a(b\mathbf{u})$.
  (ix) $1\mathbf{u} = \mathbf{u}$, where 1 is the multiplicative identity of $GF(q)$.

A subset of $V(n,q)$ is called a *subspace* of $V(n,q)$ if it is itself a vector space under the same addition and scalar multiplication as defined for $V(n,q)$.

Trivially, the set $\{\mathbf{0}\}$ and the whole space $V(n,q)$ are subspaces of $V(n,q)$. A subspace is called *non-trivial* if it contains at least one vector other than $\mathbf{0}$.

Theorem 4.1 A non-empty subset C of $V(n,q)$ is a subspace if and only if C is closed under addition and scalar multiplication, i.e. if and only if C satisfies the following two conditions:
(1) If $\mathbf{x}, \mathbf{y} \in C$, then $\mathbf{x}+\mathbf{y} \in C$.
(2) If $a \in GF(q)$ and $\mathbf{x} \in C$, then $a\mathbf{x} \in C$.

Proof It is readily verified that if C satisfies (1) and (2), then C satisfies all the axioms (i)–(ix) (with $V(n,q)$ replaced by C) for a vector space. (To show that $\mathbf{0} \in C$, choose any $\mathbf{x} \in C$; then, by (2), $\mathbf{0} = 0\mathbf{x} \in C$. Property (2) also shows that if $\mathbf{v} \in C$, then $-\mathbf{v} \in C$, for $-\mathbf{v} = (-1)\mathbf{v}$.)

Readers familiar with the theory of vector spaces over infinite fields, such as the real or complex numbers, will find that definitions and results generally carry over to the finite case, e.g. the following.

A *linear combination* of r vectors $\mathbf{v}_1, \mathbf{v}_2, \ldots, \mathbf{v}_r$ in $V(n,q)$ is a vector of the form $a_1\mathbf{v}_1 + a_2\mathbf{v}_2 + \cdots + a_r\mathbf{v}_r$, where the a_i are scalars.

It is easily verified that the set of all linear combinations of a given set of vectors of $V(n,q)$ is a subspace of $V(n,q)$.

A set of vectors $\{\mathbf{v}_1, \mathbf{v}_2, \ldots, \mathbf{v}_r\}$ is said to be *linearly dependent* if there are scalars $a_1, a_2, \ldots, a_r$, not all zero, such that
$$a_1\mathbf{v}_1 + a_2\mathbf{v}_2 + \cdots + a_r\mathbf{v}_r = \mathbf{0}.$$

A set of vectors $\{\mathbf{v}_1, \mathbf{v}_2, \ldots, \mathbf{v}_r\}$ is called *linearly independent* if it is not linearly dependent; i.e. if

$$a_1\mathbf{v}_1 + a_2\mathbf{v}_2 + \cdots + a_r\mathbf{v}_r = \mathbf{0} \Rightarrow a_1 = a_2 = \cdots = a_r = 0.$$

Let C be a subspace of $V(n, q)$. Then a subset $\{\mathbf{v}_1, \mathbf{v}_2, \ldots, \mathbf{v}_r\}$ of C is called a *generating set* (or *spanning set*) of C if every vector in C can be expressed as a linear combination of $\mathbf{v}_1, \mathbf{v}_2, \ldots, \mathbf{v}_r$.

A generating set of C which is also linearly independent is called a *basis* of C.

For example, the set

$$\{(1, 0, 0, \ldots, 0), (0, 1, 0, \ldots, 0), \ldots, (0, 0, \ldots, 0, 1)\}$$

is a basis of the whole space $V(n, q)$.

Theorem 4.2 Suppose C is a non-trivial subspace of $V(n, q)$. Then any generating set of C contains a basis of C.

Proof Suppose $\{\mathbf{v}_1, \mathbf{v}_2, \ldots, \mathbf{v}_r\}$ is a generating set of C.

If it is linearly dependent, then there are scalars $a_1, a_2, \ldots, a_r$, not all zero, such that

$$a_1\mathbf{v}_1 + a_2\mathbf{v}_2 + \cdots + a_r\mathbf{v}_r = \mathbf{0}.$$

If a_j is non-zero then

$$\mathbf{v}_j = -a_j^{-1} \sum_{i=1, i \neq j}^{r} a_i\mathbf{v}_i$$

and so $\mathbf{v}_j$ is a linear combination of the other $\mathbf{v}_i$. Thus $\mathbf{v}_j$ is redundant as a generator and can be omitted from the set $\{\mathbf{v}_1, \mathbf{v}_2, \ldots, \mathbf{v}_r\}$ to leave a smaller generating set of C. In this way we can omit redundant generators, one at a time, until we reach a linearly independent generating set. The process must end since we begin with a finite set.

Since any subspace C of $V(n, q)$ contains a finite generating set (e.g. C itself), it follows from Theorem 4.2 that every non-trivial subspace has a basis.

A basis can be thought of as a *minimal* generating set, one which does not contain any redundant generators.

Theorem 4.3 Suppose $\{\mathbf{v}_1, \mathbf{v}_2, \ldots, \mathbf{v}_k\}$ is a basis of a subspace C

of $V(n, q)$. Then
(i) every vector of C can be expressed *uniquely* as a linear combination of the basis vectors.
(ii) C contains exactly q^k vectors.

Proof (i) Suppose a vector $\mathbf{x}$ of C is represented in two ways as a linear combination of $\mathbf{v}_1, \mathbf{v}_2, \ldots, \mathbf{v}_k$. That is,

$$\mathbf{x} = a_1\mathbf{v}_1 + a_2\mathbf{v}_2 + \cdots + a_k\mathbf{v}_k$$
$$\text{and} \quad \mathbf{x} = b_1\mathbf{v}_1 + b_2\mathbf{v}_2 + \cdots + b_k\mathbf{v}_k.$$

Then $(a_1 - b_1)\mathbf{v}_1 + (a_2 - b_2)\mathbf{v}_2 + \cdots + (a_k - b_k)\mathbf{v}_k = \mathbf{0}$. But the set $\{\mathbf{v}_1, \mathbf{v}_2, \ldots, \mathbf{v}_k\}$ is linearly independent and so $a_i - b_i = 0$ for $i = 1, 2, \ldots, k$; i.e. $a_i = b_i$ for $i = 1, 2, \ldots, k$.
(ii) By (i), the q^k vectors $\sum_{i=1}^{k} a_i\mathbf{v}_i$ ($a_i \in GF(q)$) are precisely the distinct vectors of C.

It follows from Theorem 4.3 that any two bases of a subspace C contain the same number k of vectors, where $|C| = q^k$, and this number k is called the *dimension* of the subspace C; it is denoted by $\dim(C)$.
We have already exhibited a basis of $V(n, q)$ having n vectors and so $\dim(V(n, q)) = n$.

Exercises 4

4.1 Show that a non-empty subset C of $V(n, q)$ is a subspace if and only if $a\mathbf{x} + b\mathbf{y} \in C$ for all $a, b \in GF(q)$ and for all $\mathbf{x}, \mathbf{y} \in C$.

4.2 Show that the set E_n of all even-weight vectors of $V(n, 2)$ is a subspace of $V(n, 2)$. What is the dimension of E_n? [*Hint:* See Exercise 2.4.] Write down a basis for E_n.

4.3 Let C be the subspace of $V(4, 3)$ having as generating set $\{(0, 1, 2, 1), (1, 0, 2, 2), (1, 2, 0, 1)\}$. Find a basis of C. What is $\dim(C)$?

4.4 Let $\mathbf{u}$ and $\mathbf{v}$ be vectors in $V(n, q)$. Show that the set $\{\mathbf{u}, \mathbf{v}\}$ is linearly independent if and only if $\mathbf{u}$ and $\mathbf{v}$ are non-zero and $\mathbf{v}$ is not a scalar multiple of $\mathbf{u}$.

4.5 Suppose $\{\mathbf{x}_1, \mathbf{x}_2, \ldots, \mathbf{x}_k\}$ is a basis for a subspace C of $V(n, q)$. Show that we get a basis for the same subspace C

if we either

(a)   replace an $\mathbf{x}_i$ by a non-zero scalar multiple of itself, or

(b)   replace an $\mathbf{x}_i$ by $\mathbf{x}_i + a\mathbf{x}_j$, for some scalar a, with $j \neq i$.

4.6   Suppose F is a field of characteristic p. Show that F can be regarded as a vector space over $GF(p)$. Deduce that any finite field has order equal to a power of some prime number.

4.7   From the vector space $V(3, q)$, an incidence structure P_q is defined as follows.

The 'points' of P_q are the one-dimensional subspaces of $V(3, q)$. The 'lines' of P_q are the two-dimensional subspaces of $V(3, q)$. The point P 'belongs to' the line L if and only if P is a subspace of L.

Prove that P_q is a finite projective plane of order q. List the points and lines of P_2 and check that it has the same structure as the seven-point plane defined in Example 2.19.

5 Introduction to linear codes

Throughout this chapter, we assume that the alphabet F_q is the Galois field $GF(q)$, where q is a prime power, and we regard $(F_q)^n$ as the vector space $V(n, q)$. A vector $(x_1, x_2, \ldots, x_n)$ will usually be written simply as $x_1 x_2 \cdots x_n$.

A *linear code* over $GF(q)$ is just a subspace of $V(n, q)$, for some positive integer n.

Thus a subset C of $V(n, q)$ is a linear code if and only if
(1) $\mathbf{u} + \mathbf{v} \in C$, for all $\mathbf{u}$ and $\mathbf{v}$ in C, and
(2) $a\mathbf{u} \in C$, for all $\mathbf{u} \in C$, $a \in GF(q)$.

In particular, a binary code is linear if and only if the sum of any two codewords is a codeword. It is easily checked that the codes C_1, C_2 and C_3 of Example 1.5, and the code C of Example 2.23, are all linear.

If C is a k-dimensional subspace of $V(n, q)$, then the linear code C is called an $[n, k]$-*code*, or sometimes, if we wish to specify also the minimum distance d of C, an $[n, k, d]$-*code*.

Notes (i) A q-ary $[n, k, d]$-code is also a q-ary (n, q^k, d)-code (by Theorem 4.3), but, of course, not every (n, q^k, d)-code is an $[n, k, d]$-code.

(ii) The all-zero vector $\mathbf{0}$ automatically belongs to a linear code.

(iii) Some authors have referred to linear codes as 'group codes'.

The *weight* $w(\mathbf{x})$ of a vector $\mathbf{x}$ in $V(n, q)$ is defined to be the number of non-zero entries of $\mathbf{x}$. One of the most useful properties of a linear code is that its minimum distance is equal to the smallest of the weights of the non-zero codewords. To prove this we need a simple lemma.

Lemma 5.1 If $\mathbf{x}$ and $\mathbf{y} \in V(n, q)$, then

$$d(\mathbf{x}, \mathbf{y}) = w(\mathbf{x} - \mathbf{y}).$$

Proof The vector $\mathbf{x} - \mathbf{y}$ has non-zero entries in precisely those places where $\mathbf{x}$ and $\mathbf{y}$ differ.

Remark For $q = 2$, Lemma 5.1 is the same as Lemma 2.5, bearing in mind that 'plus' is the same as 'minus' when working modulo 2.

Theorem 5.2 Let C be a linear code and let $w(C)$ be the smallest of the weights of the non-zero codewords of C. Then

$$d(C) = w(C).$$

Proof There exist codewords $\mathbf{x}$ and $\mathbf{y}$ of C such that $d(C) = d(\mathbf{x}, \mathbf{y})$. Then, by Lemma 5.1,

$$d(C) = w(\mathbf{x} - \mathbf{y}) \geqslant w(C),$$

since $\mathbf{x} - \mathbf{y}$ is a codeword of the linear code C.
   On the other hand, for some codeword $\mathbf{x} \in C$,

$$w(C) = w(\mathbf{x}) = d(\mathbf{x}, \mathbf{0}) \geqslant d(C),$$

since $\mathbf{0}$ belongs to the linear code C. Hence $d(C) \geqslant w(C)$ and $w(C) \geqslant d(C)$, giving $d(C) = w(C)$.

   We now list some of the advantages and disadvantages of restricting one's attention to linear codes.

Advantage 1 For a general code with M codewords, to find the minimum distance we might have to make $\binom{M}{2} = \frac{1}{2}M(M - 1)$ comparisons (as in Example 2.23). However, Theorem 5.2 enables the minimum distance of a *linear* code to be found by examining only the weights of the $M - 1$ non-zero codewords.
   Note how much easier it is now to show that the code of Example 2.23 has minimum distance 3, if we know that it is linear.

Advantage 2 To specify a non-linear code, we may have to list all the codewords. We can specify a linear $[n, k]$-code by simply giving a basis of k codewords.

Definition A $k \times n$ matrix whose rows form a basis of a linear $[n, k]$-code is called a *generator matrix* of the code.

Examples 5.3 (i) The code C_2 of Example 1.5 is a $[3, 2, 2]$-code with generator matrix $\begin{bmatrix} 0 & 1 & 1 \\ 1 & 0 & 1 \end{bmatrix}$.

(ii) The code C of Example 2.23 is a $[7, 4, 3]$-code with generator matrix

$$\begin{bmatrix} 1 & 1 & 1 & 1 & 1 & 1 & 1 \\ 1 & 0 & 0 & 0 & 1 & 0 & 1 \\ 1 & 1 & 0 & 0 & 0 & 1 & 0 \\ 0 & 1 & 1 & 0 & 0 & 0 & 1 \end{bmatrix}$$

(iii) The q-ary repetition code of length n over $GF(q)$ is an $[n, 1, n]$-code with generator matrix

$$[1 \ 1 \cdots 1].$$

Advantage 3 There are nice procedures for encoding and decoding a linear code (See Chapters 6 and 7).

Disadvantage 1 Linear q-ary codes are not defined unless q is a prime power. However, reasonable q-ary codes, for q not a prime power, can often be obtained from linear codes over a larger alphabet. For example, we shall see in Chapter 7 how good decimal (i.e. 10-ary) codes can be obtained from linear 11-ary codes by omitting all codewords containing a given fixed symbol. This idea has already been illustrated in Chapter 3, for the ISBN code can be obtained in such a way from the linear 11-ary code

$$\left\{ x_1 x_2 \cdots x_{10} \in V(10, 11): \sum_{i=1}^{10} i x_i = 0 \right\}.$$

Disadvantage 2 The restriction to linear codes might be a restriction to weaker codes than desired. However, it turns out that codes which are optimal in some way are very frequently linear. For example, for every set of parameters for which it is known that there exists a non-trivial perfect code, there exists a perfect *linear* code with those parameters. Notice also how often the value of $A_2(n, d)$ in Table 2.4 is a power of 2. It is usually, though not always, the case that such a value of $A_2(n, d)$ is achieved by a linear code.

Equivalence of linear codes

The definition of equivalence of codes given in Chapter 2 is modified for linear codes, by allowing only those permutations of symbols which are given by multiplication by a non-zero scalar. Thus two linear codes over $GF(q)$ are called *equivalent* if one can be obtained from the other by a combination of operations of the following types.

(A) permutation of the positions of the code;
(B) multiplication of the symbols appearing in a fixed position by a non-zero scalar.

Theorem 5.4 Two $k \times n$ matrices generate equivalent linear $[n, k]$-codes over $GF(q)$ if one matrix can be obtained from the other by a sequence of operations of the following types:

(R1) Permutation of the rows.
(R2) Multiplication of a row by a non-zero scalar.
(R3) Addition of a scalar multiple of one row to another.
(C1) Permutation of the columns.
(C2) Multiplication of any column by a non-zero scalar.

Proof The row operations (R1), (R2) and (R3) preserve the linear independence of the rows of a generator matrix and simply replace one basis by another of the same code (see Exercise 4.5). Operations of type (C1) and (C2) convert a generator matrix to one for an equivalent code.

Theorem 5.5 Let G be a generator matrix of an $[n, k]$-code. Then by performing operations of types (R1), (R2), (R3), (C1) and (C2), G can be transformed to the *standard form*

$$[I_k \,|\, A],$$

where I_k is the $k \times k$ identity matrix, and A is a $k \times (n - k)$ matrix.

Proof During a sequence of transformations of the matrix G, we denote by g_{ij} the (i, j)th entry of the matrix under consideration *at the time* and by $r_1, r_2, \ldots, r_k$ and $c_1, c_2, \ldots, c_n$ the rows and columns respectively of this matrix.

The following three-step procedure is applied for $j = 1$, $2, \ldots, k$ in turn, the jth application transforming column c_j into its desired form (with 1 in the jth position and 0s elsewhere), leaving unchanged the first $j - 1$ columns already suitably transformed. Suppose then that G has already been transformed to

$$\begin{bmatrix} 1 & 0 & \cdots & 0 & g_{1j} & \cdots & g_{1n} \\ 0 & 1 & \cdots & 0 & g_{2j} & \cdots & g_{2n} \\ \vdots & \vdots & \ddots & \vdots & \vdots & & \vdots \\ 0 & 0 & \cdots & 1 & g_{j-1,j} & \cdots & g_{j-1,n} \\ 0 & 0 & \cdots & 0 & g_{jj} & \cdots & g_{jn} \\ \vdots & \vdots & & \vdots & \vdots & & \vdots \\ 0 & 0 & \cdots & 0 & g_{kj} & \cdots & g_{kn} \end{bmatrix}.$$

Step 1 If $g_{jj} \neq 0$, go to Step 2. If $g_{jj} = 0$, and if for some $i > j, g_{ij} \neq 0$, then interchange r_j and r_i. If $g_{jj} = 0$ and $g_{ij} = 0$ for all $i > j$, then choose h such that $g_{jh} \neq 0$ and interchange c_j and c_h.
Step 2 We now have $g_{jj} \neq 0$. Multiply r_j by g_{jj}^{-1}.
Step 3 We now have $g_{jj} = 1$. For each of $i = 1, 2, \ldots, k$, with $i \neq j$, replace r_i by $r_i - g_{ij} r_j$.
The column c_j now has the desired form.

After this procedure has been applied for $j = 1, 2, \ldots, k$, the generator matrix will have standard form.

Notes (1) If G can be transformed into a standard form matrix G' by row operations only (this will be the case if and only if the first k columns of G are linearly independent), then G' will actually generate the *same* code as does G. But if operations (C1) and (C2) are also used, then G' will generate a code which is equivalent to, though not necessarily the same as, that generated by G. The procedure described in the preceding proof is designed to give a standard form generator matrix for the *same* code whenever this is possible.

(2) In practice, inspection of the generator matrix G will

often suggest a quicker way to transform to standard form, as in Example 5.6(iii) below.

(3) The standard form $[I_k \mid A]$ of a generator matrix is not unique; for example, permutation of the columns of A will give a generator matrix for an equivalent code.

Examples 5.6 (i) See Example 5.3(i). Interchanging rows gives the standard form generator matrix

$$\begin{bmatrix} 1 & 0 & 1 \\ 0 & 1 & 1 \end{bmatrix}$$

for the code C_2.

(ii) We will use the procedure of Theorem 5.5 to transform the generator matrix of Example 5.3(ii) to standard form.

$$\begin{bmatrix} 1 & 1 & 1 & 1 & 1 & 1 & 1 \\ 1 & 0 & 0 & 0 & 1 & 0 & 1 \\ 1 & 1 & 0 & 0 & 0 & 1 & 0 \\ 0 & 1 & 1 & 0 & 0 & 0 & 1 \end{bmatrix} \xrightarrow[\substack{r_2 \to r_2 - r_1 \\ r_3 \to r_3 - r_1}]{} \begin{bmatrix} 1 & 1 & 1 & 1 & 1 & 1 & 1 \\ 0 & 1 & 1 & 1 & 0 & 1 & 0 \\ 0 & 0 & 1 & 1 & 1 & 0 & 1 \\ 0 & 1 & 1 & 0 & 0 & 0 & 1 \end{bmatrix}$$

$$\xrightarrow[\substack{r_1 \to r_1 - r_2 \\ r_4 \to r_4 - r_2}]{} \begin{bmatrix} 1 & 0 & 0 & 0 & 1 & 0 & 1 \\ 0 & 1 & 1 & 1 & 0 & 1 & 0 \\ 0 & 0 & 1 & 1 & 1 & 0 & 1 \\ 0 & 0 & 0 & 1 & 0 & 1 & 1 \end{bmatrix}$$

$$\xrightarrow[r_2 \to r_2 - r_3]{} \begin{bmatrix} 1 & 0 & 0 & 0 & 1 & 0 & 1 \\ 0 & 1 & 0 & 0 & 1 & 1 & 1 \\ 0 & 0 & 1 & 1 & 1 & 0 & 1 \\ 0 & 0 & 0 & 1 & 0 & 1 & 1 \end{bmatrix}$$

$$\xrightarrow[r_3 \to r_3 - r_4]{} \begin{bmatrix} 1 & 0 & 0 & 0 & 1 & 0 & 1 \\ 0 & 1 & 0 & 0 & 1 & 1 & 1 \\ 0 & 0 & 1 & 0 & 1 & 1 & 0 \\ 0 & 0 & 0 & 1 & 0 & 1 & 1 \end{bmatrix}$$

(iii) Consider the $[6, 3]$-code over $GF(3)$ having generator matrix

$$\begin{bmatrix} 0 & 0 & 0 & 1 & 1 & 1 \\ 0 & 1 & 1 & 0 & 1 & 2 \\ 1 & 0 & 2 & 0 & 1 & 1 \end{bmatrix}.$$

An obvious permutation of the columns gives the standard form
generator matrix

$$\begin{bmatrix} 1 & 0 & 0 & 0 & 1 & 1 \\ 0 & 1 & 0 & 1 & 1 & 2 \\ 0 & 0 & 1 & 2 & 1 & 1 \end{bmatrix}$$

for an equivalent code.

Exercises 5

5.1 Is the binary $(11, 24, 5)$-code of Exercise 2.12 linear?
(There is no need to examine any codewords).

5.2 Exercise 4.2 shows that E_n, the code of all even-weight
vectors of $V(n, 2)$, is linear. What are the parameters
$[n, k, d]$ of E_n? Write down a generator matrix for E_n in
standard form.

5.3 Let H be an $r \times n$ matrix over $GF(q)$. Prove that the set
$C = \{\mathbf{x} \in V(n, q) \mid \mathbf{x}H^T = \mathbf{0}\}$ is a linear code. [*Remark:* we
shall show in Chapter 7 that *every* linear code may be
defined by means of such a matrix H, which is called a
parity-check matrix of the code.]

5.4 (i) Show that if C is a binary linear code, then the code
obtained by adding an overall parity check to C is
also linear.
(ii) Find a generator matrix for a binary $[8, 4, 4]$-code.

5.5 Prove that, in a binary linear code, either all the codewords have even weight or exactly half have even weight
and half have odd weight.

5.6 Let C_1 and C_2 be binary linear codes having the generator
matrices

$$G_1 = \begin{bmatrix} 1 & 1 & 1 & 1 & 0 \\ 0 & 0 & 1 & 1 & 1 \end{bmatrix} \quad \text{and} \quad G_2 = \begin{bmatrix} 1 & 0 & 0 & 1 & 1 & 0 & 1 \\ 0 & 1 & 0 & 1 & 0 & 1 & 1 \\ 0 & 0 & 1 & 0 & 1 & 1 & 1 \end{bmatrix}.$$

List the codewords of C_1 and C_2 and hence find the
minimum distance of each code. (Use Theorem 5.2.)

5.7 Let C be the ternary linear code with generator matrix

$$\begin{bmatrix} 1 & 0 & 1 & 1 \\ 0 & 1 & 1 & 2 \end{bmatrix}.$$

List the codewords of C and use Theorem 5.2 to find the minimum distance of C. Deduce that C is a perfect code.

5.8 Let $B_q(n, d)$ denote the largest value of M for which there exists a *linear* q-ary (n, M, d)-code (q is a prime power). Clearly the value of $B_q(n, d)$ is less than or equal to the value of $A_q(n, d)$, which was defined in Chapter 2. Determine the values of $B_2(8, 3)$, $B_2(8, 4)$ and $B_2(8, 5)$. Is it true that $B_2(n, d) = A_2(n, d)$ for each of these cases?

5.9 Exercise 2.3 shows that $A_q(3, 2) = q^2$ for any integer $q \geq 2$. Show that, if q is a prime power, then $B_q(3, 2) = q^2$.

5.10 Suppose $[I_k \mid A]$ is a standard form generator matrix for a linear code C. Show that any permutation of the rows of A gives a generator matrix for a code which is equivalent to C.

5.11 Let C be the binary linear code with generator matrix

$$\begin{bmatrix} 1 & 1 & 1 & 0 & 0 & 0 & 0 \\ 1 & 0 & 0 & 1 & 1 & 0 & 0 \\ 1 & 0 & 0 & 0 & 0 & 1 & 1 \\ 0 & 1 & 0 & 1 & 0 & 1 & 0 \end{bmatrix}$$

Find a generator matrix for C in standard form. Is C the same as the code of Example 5.6(ii)? Is C equivalent to the code of Example 5.6(ii)?

5.12 Suppose C_1 and C_2 are binary linear codes. Let C_3 be the code given by the $(u \mid u + v)$ construction of Exercise 2.17. Show that C_3 is linear.
Deduce that $B_2(2d, d) = 4d$ when d is a power of 2.

6 Encoding and decoding with a linear code

Encoding with a linear code

Let C be an $[n, k]$-code over $GF(q)$ with generator matrix G. C contains q^k codewords and so can be used to communicate any one of q^k distinct messages. We identify these messages with the q^k k-tuples of $V(k, q)$ and we *encode* a message vector $\mathbf{u} = u_1 u_2 \cdots u_k$ simply by multiplying it on the right by G. If the rows of G are $\mathbf{r}_1, \mathbf{r}_2, \ldots, \mathbf{r}_k$, then

$$\mathbf{u}G = \sum_{i=1}^{k} u_i \mathbf{r}_i$$

and so $\mathbf{u}G$ is indeed a codeword of C, being a linear combination of the rows of the generator matrix. Note that the encoding function $\mathbf{u} \rightarrow \mathbf{u}G$ maps the vector space $V(k, q)$ on to a k-dimensional subspace (namely the code C) of $V(n, q)$.

The encoding rule is even simpler if G is in standard form. Suppose $G = [I_k \,|\, A]$, where $A = [a_{ij}]$ is a $k \times (n - k)$ matrix. Then the message vector $\mathbf{u}$ is encoded as

$$\mathbf{x} = \mathbf{u}G = x_1 x_2 \cdots x_k x_{k+1} \cdots x_n,$$

where $x_i = u_i$, $1 \leqslant i \leqslant k$, are the *message digits*

and

$$x_{k+i} = \sum_{j=1}^{k} a_{ji} u_j, \qquad 1 \leqslant i \leqslant n - k,$$

are the *check digits*. The check digits represent *redundancy* which has been added to the message to give protection against noise.

Example 6.1 Let C be the binary $[7, 4]$-code of Example 5.3(ii), for which we found in Example 5.6(ii) the standard form generator matrix

$$G = \begin{bmatrix} 1 & 0 & 0 & 0 & 1 & 0 & 1 \\ 0 & 1 & 0 & 0 & 1 & 1 & 1 \\ 0 & 0 & 1 & 0 & 1 & 1 & 0 \\ 0 & 0 & 0 & 1 & 0 & 1 & 1 \end{bmatrix}.$$

A message vector (u_1, u_2, u_3, u_4) is encoded as

$$(u_1, u_2, u_3, u_4, u_1 + u_2 + u_3, u_2 + u_3 + u_4, u_1 + u_2 + u_4).$$

For example,

0 0 0 0	is encoded as			0 0 0 0 0 0 0,
1 0 0 0	,,	,,	,,	1 0 0 0 1 0 1,
1 1 1 0	,,	,,	,,	1 1 1 0 1 0 0.

For a general linear code, we summarize the encoding part of the communication scheme (see Fig. 1.1) in Fig. 6.2.

Fig. 6.2

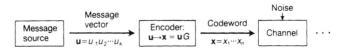

Decoding with a linear code

Suppose the codeword $\mathbf{x} = x_1 x_2 \cdots x_n$ is sent through the channel and that the received vector is $\mathbf{y} = y_1 y_2 \cdots y_n$. We define the *error vector* $\mathbf{e}$ to be

$$\mathbf{e} = \mathbf{y} - \mathbf{x} = e_1 e_2 \cdots e_n.$$

The decoder must decide from $\mathbf{y}$ which codeword $\mathbf{x}$ was transmitted, or equivalently which error vector $\mathbf{e}$ has occurred. An elegant nearest neighbour decoding scheme for linear codes, devised by Slepian (1960), uses the fact that a linear code is a subgroup of the additive group $V(n, q)$. The reader who is not familiar with elementary group theory should not be deterred as we shall not be assuming any prior knowledge of the subject here.

Definition Suppose that C is an $[n, k]$-code over $GF(q)$ and that $\mathbf{a}$ is any vector in $V(n, q)$. Then the set $\mathbf{a} + C$ defined by

$$\mathbf{a} + C = \{\mathbf{a} + \mathbf{x} \mid \mathbf{x} \in C\}$$

is called a *coset* of C.

Lemma 6.3 Suppose that $\mathbf{a} + C$ is a coset of C and that

$\mathbf{b} \in \mathbf{a} + C$. Then
$$\mathbf{b} + C = \mathbf{a} + C.$$

Proof Since $\mathbf{b} \in \mathbf{a} + C$, we have $\mathbf{b} = \mathbf{a} + \mathbf{x}$, for some $\mathbf{x} \in C$. Now if $\mathbf{b} + \mathbf{y} \in \mathbf{b} + C$, then

$$\mathbf{b} + \mathbf{y} = (\mathbf{a} + \mathbf{x}) + \mathbf{y} = \mathbf{a} + (\mathbf{x} + \mathbf{y}) \in \mathbf{a} + C.$$

Hence $\mathbf{b} + C \subseteq \mathbf{a} + C$. On the other hand, if $\mathbf{a} + \mathbf{z} \in \mathbf{a} + C$, then

$$\mathbf{a} + \mathbf{z} = (\mathbf{b} - \mathbf{x}) + \mathbf{z} = \mathbf{b} + (\mathbf{z} - \mathbf{x}) \in \mathbf{b} + C.$$

Hence $\mathbf{a} + C \subseteq \mathbf{b} + C$, and so $\mathbf{b} + C = \mathbf{a} + C$.

The following theorem is a particular case of Lagrange's well-known theorem for subgroups.

Theorem 6.4 (Lagrange) Suppose C is an $[n, k]$-code over $GF(q)$. Then

(i)   every vector of $V(n, q)$ is in some coset of C,
(ii) every coset contains exactly q^k vectors,
(iii) two cosets either are disjoint or coincide (partial overlap is impossible).

Proof (i)   If $\mathbf{a} \in V(n, q)$, then $\mathbf{a} = \mathbf{a} + \mathbf{0} \in \mathbf{a} + C$.
(ii) The mapping from C to $\mathbf{a} + C$ defined by

$$\mathbf{x} \to \mathbf{a} + \mathbf{x},$$

for all $\mathbf{x} \in C$, is easily shown to be one-to-one. Hence $|\mathbf{a} + C| = |C| = q^k$.

(iii) Suppose the cosets $\mathbf{a} + C$ and $\mathbf{b} + C$ overlap. Then for some vector $\mathbf{v}$, we have $\mathbf{v} \in (\mathbf{a} + C) \cap (\mathbf{b} + C)$. Thus, for some $\mathbf{x}, \mathbf{y} \in C$,

$$\mathbf{v} = \mathbf{a} + \mathbf{x} = \mathbf{b} + \mathbf{y}.$$

Hence $\mathbf{b} = \mathbf{a} + (\mathbf{x} - \mathbf{y}) \in \mathbf{a} + C$, and so by Lemma 6.3, $\mathbf{b} + C = \mathbf{a} + C$.

Example 6.5 Let C be the binary $[4, 2]$-code with generator matrix

$$G = \begin{bmatrix} 1 & 0 & 1 & 1 \\ 0 & 1 & 0 & 1 \end{bmatrix}.$$

i.e. $C = \{0000, 1011, 0101, 1110\}$.

Then the cosets of C are

$$0000 + C = C \text{ itself,}$$
$$1000 + C = \{1000, 0011, 1101, 0110\},$$
$$0100 + C = \{0100, 1111, 0001, 1010\},$$
and $\qquad 0010 + C = \{0010, 1001, 0111, 1100\}.$

Note that the coset $0001 + C$ is $\{0001, 1010, 0100, 1111\}$, which is the same as the coset $0100 + C$. This could have been predicted from Lemma 6.3, since $0001 \in 0100 + C$. Similarly we must have, for example, $0111 + C = 0010 + C$.

Definition The vector having minimum weight in a coset is called the *coset leader*. (If there is more than one vector with the minimum weight, we choose one at random and call it the coset leader. For example, in Example 6.5, 0001 is an alternative coset leader to 0100 for the coset $0100 + C$).

Theorem 6.4 shows that $V(n, q)$ is partitioned into disjoint cosets of C:

$$V(n, q) = (\mathbf{0} + C) \cup (\mathbf{a}_1 + C) \cup \cdots \cup (\mathbf{a}_s + C),$$

where $s = q^{n-k} - 1$, and, by Lemma 6.3, we may take $\mathbf{0}, \mathbf{a}_1, \ldots, \mathbf{a}_s$ to be the coset leaders.

A *(Slepian) standard array* for an $[n, k]$-code C is a $q^{n-k} \times q^k$ array of all the vectors in $V(n, q)$ in which the first row consists of the code C with $\mathbf{0}$ on the extreme left, and the other rows are the cosets $\mathbf{a}_i + C$, each arranged in corresponding order, with the coset leader on the left. A standard array may be constructed as follows:

Step 1 List the codewords of C, starting with $\mathbf{0}$, as the first row.
Step 2 Choose any vector $\mathbf{a}_1$, not in the first row, of minimum weight. List the coset $\mathbf{a}_1 + C$ as the second row by putting $\mathbf{a}_1$ under $\mathbf{0}$ and $\mathbf{a}_1 + \mathbf{x}$ under $\mathbf{x}$ for each $\mathbf{x} \in C$.
Step 3 From those vectors not in rows 1 and 2, choose $\mathbf{a}_2$ of minimum weight and list the coset $\mathbf{a}_2 + C$ as in Step 2 to get the third row.
Step 4 Continue in this way until all the cosets are listed and every vector of $V(n, q)$ appears exactly once.

Example 6.6 A standard array for the code of Example 6.5 is

codewords	→ 0000	1011	0101	1110
	1000	0011	1101	0110
	0100	1111	0001	1010
	0010	1001	0111	1100

↑
coset leaders

Note that in a standard array, each entry is the sum of the codeword at the top of its column and the coset leader at the extreme left of its row. We now describe how the decoder uses the standard array.

When **y** is received (e.g. 1111 in the above example), its position in the array is found. Then the decoder decides that the error vector **e** is the coset leader (0100) found at the extreme left of **y** and **y** is decoded as the codeword $\mathbf{x} = \mathbf{y} - \mathbf{e}$ (1011) at the top of the column containing **y**.

Briefly, a received vector is decoded as the codeword at the top of its column in the standard array.

The error vectors which will be corrected are precisely the coset leaders, irrespective of which codeword is transmitted. By choosing a minimum weight vector in each coset as coset leader, we ensure that standard array decoding is a nearest neighbour decoding scheme.

In Example 6.6, with the given array, a single error will be corrected if it occurs in any of the first 3 places (e.g. (a) below) but not if it occurs in the 4th place (e.g. (b) below).

	Message		Codeword		Channel + noise		Received vector		Decoded word		Received message
(a)	01	→	0101	→	0101	→	0001	→	0101	→	01
(b)	01	→	0101	→	0101	→	0100	→	0000	→	00

Notes (1) In practice, the above decoding scheme is too slow for large codes and also too costly in terms of storage requirements. A more sophisticated way of carrying out standard array decoding, known as 'syndrome decoding', will be described in Chapter 7.

(2) In Example (b) above, the message symbols 01 were

actually unaffected by noise and yet, after decoding, the wrong message 00 was received. This is an instance of more harm than good ensuing from the addition of redundancy. But in order to get a sensible measure of how good a code is, we must calculate the *probability* that a received vector will be decoded as the codeword which was sent. Since the error vectors which will be corrected by standard array decoding are the same whichever codeword is sent, this calculation is extremely easy for a linear code, as we now show.

Probability of error correction

For simplicity, we restrict our attention for the remainder of this chapter to *binary* linear codes. We assume that the channel is binary symmetric with symbol error probability p. We saw in Chapter 1 that the probability that the error vector is a given vector of weight i is $p^i(1-p)^{n-i}$ and so the following theorem follows immediately.

Theorem 6.7 Let C be a binary $[n, k]$-code, and for $i = 0$, $1, \ldots, n$ let α_i denote the number of coset leaders of weight i. Then the probability $P_{\text{corr}}(C)$ that a received vector decoded by means of a standard array is the codeword which was sent is given by

$$P_{\text{corr}}(C) = \sum_{i=0}^{n} \alpha_i p^i (1-p)^{n-i}.$$

Example 6.8 For the $[4, 2]$-code of Example 6.6, the coset leaders are 0000, 1000, 0100 and 0010. Hence $\alpha_0 = 1$, $\alpha_1 = 3$, $\alpha_2 = \alpha_3 = \alpha_4 = 0$, and so

$$P_{\text{corr}}(C) = (1-p)^4 + 3p(1-p)^3$$
$$= (1-p)^3(1+2p).$$

If $p = 0.01$, then $P_{\text{corr}}(C) = 0.9897$. The probability that a decoded word is *not* the word sent, i.e. the *word error rate*, is

$$P_{\text{err}}(C) = 1 - P_{\text{corr}}(C),$$

which, for $p = 0.01$, is 0.0103.

Without coding, the probability of a 2-digit message being received incorrectly is $1 - (1-p)^2$ which, for $p = 0.01$, is 0.0199.

So, for $p = 0.01$, we have nearly halved the word error rate at the expense of having to send two check symbols with every 2-digit message.

Remark 6.9 If $d(C) = 2t + 1$ or $2t + 2$, then C can correct any t errors. Hence every vector of weight $\leq t$ is a coset leader and so $\alpha_i = \binom{n}{i}$ for $0 \leq i \leq t$. But for $i > t$, the α_i can be extremely difficult to calculate and are unknown even for some very well-known families of codes. One case for which there is no such difficulty is that of perfect codes; since the error vectors corrected by a perfect $[n, k, 2t + 1]$-code are precisely those vectors of weight $\leq t$, we have $\alpha_i = \binom{n}{i}$ for $0 \leq i \leq t$ and $\alpha_i = 0$ for $i > t$.

A linear $[n, k]$-code C uses n symbols to send k message symbols. It is said to have *rate* $R(C) = k/n$. Thus the rate of a code is the ratio of the number of message symbols to the total number of symbols sent and so a good code will have a high rate.

Example 6.10 Let us return to Example 1.5 and consider how a route can most accurately be communicated if we impose the condition that the rate of the code used must be at least $\frac{1}{2}$, i.e. that there is time enough to send only as many check symbols as there are message symbols. We will assume the channel to be binary symmetric with $p = 0.01$.

It might at first appear that we can do no better than to use the $[4, 2]$-code of Example 6.6, for which we found in Example 6.8 that $P_{err} = 0.0103$. It is not hard to see that this is the best we can do if we limit ourselves to using just four codewords, one for each possible message N, W, E or S. But consider the following strategy.

We first identify N, W, E and S with the message vectors 00, 01, 10 and 11 and convert the route (e.g. NNWN $\cdots$) to a long string of message symbols (00000100 $\cdots$). We then break the string into blocks of 4 and encode each block into a length 7 codeword by means of the $[7, 4]$-code C considered in Examples 2.23, 5.6 and 6.1. By Remark 6.9, since C is a perfect $[7, 4, 3]$-code, we have $\alpha_0 = 1$, $\alpha_1 = 7$ and $\alpha_i = 0$ for $i > 1$. (Note that there is no need to construct a standard array to find the α_i

in this case.) Hence

$$P_{err}(C) = 1 - (1-p)^7 - 7p(1-p)^6$$
$$\simeq 0.002 \quad \text{if } p = 0.01.$$

Thus the number of codewords (and hence messages) received in error after decoding with this [7, 4]-code is about one-fifth of the number received in error when using the best [4, 2]-code. And yet we are sending the information at a more efficient rate, for $R(C) = \frac{4}{7} > \frac{1}{2}$.

One lesson to be learned from this example is that if we first represent our information by a long string of binary digits, we need not be too restricted in our choice of [n, k]-code, for we can just encode the message symbols k at a time. We shall see in Exercise 6.6 that by using a [23, 12]-code, which has rate $>\frac{1}{2}$, we can get the word error rate P_{err} down to approximately 0.000 08.

It is beginning to look as though we can make the word error rate as small as we wish by using a long enough code (but still having rate $\geq\frac{1}{2}$). Indeed it is a consequence of the following remarkable theorem of Shannon (1948) that, for a binary symmetric channel with symbol error probability p, we can communicate at a given rate R with as small a word error rate as we wish, provided R is less than a certain function of p called the capacity of the channel.

Definition The *capacity* $\mathscr{C}(p)$ of a binary symmetric channel with symbol error probability p is

$$\mathscr{C}(p) = 1 + p \log_2 p + (1-p) \log_2 (1-p).$$

Fig. 6.11

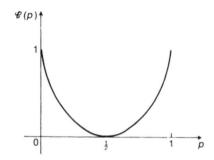

Theorem 6.12 (Shannon's theorem; proof omitted)   Suppose a

channel is binary symmetric with symbol error probability p. Suppose R is a number satisfying $R < \mathscr{C}(p)$. Then for any $\varepsilon > 0$, there exists, for sufficiently large n, an $[n, k]$-code C of rate $k/n \geqslant R$ such that $P_{\text{err}}(C) < \varepsilon$.

(A similar result holds for non-binary codes, but with a different definition of capacity).

The proof of this result may be found in van Lint (1982) or McEliece (1977). Unfortunately, the theorem has so far been proved only by probabilistic methods and does not tell us how to construct such codes. It should be borne in mind also that for practical purposes we require codes which are easily encoded and decoded and that this is less likely to be the case for long codes with many codewords.

Example 6.13 It may be calculated that $\mathscr{C}(0.01) \cong 0.92$. Thus, for $p = 0.01$, even if we insist on transmitting at a rate of $\frac{9}{10}$, we can, in theory, make P_{err} as small as we wish by making n (and k) sufficiently large.

Symbol error rate

Since some of the message symbols may be correct even if the decoder outputs the wrong codeword, a more useful quantity might be the *symbol error rate* P_{symb}, the average probability that a message symbol is in error after decoding. A method for calculating P_{symb} is given in Exercise 6.7, but it is more difficult to calculate than P_{err} and is not known for many codes. Note also that the result of Exercise 6.9 shows that Shannon's theorem remains true if we replace P_{err} by P_{symb}.

Probability of error detection

Suppose now that a binary linear code is to be used only for error detection. The decoder will fail to detect errors which have occurred if and only if the received vector $\mathbf{y}$ is a codeword different from the codeword $\mathbf{x}$ which was sent, i.e. if and only if the error vector $\mathbf{e} = \mathbf{y} - \mathbf{x}$ is itself a non-zero codeword (since C is linear). Thus the probability $P_{\text{undetec}}(C)$ that an incorrect codeword will be received is independent of the codeword sent and is given by the following theorem.

Theorem 6.14 Let C be a binary $[n, k]$-code and let A_i denote the number of codewords of C of weight i. Then, if C is used for error detection, the probability of an incorrect message being received undetected is

$$P_{\text{undetec}}(C) = \sum_{i=1}^{n} A_i p^i (1-p)^{n-i}.$$

(Note that, unlike the formula of Theorem 6.7 for $P_{\text{corr}}(C)$, the summation here starts at $i = 1$).

Example 6.15 With the code of Example 6.6,

$$P_{\text{undetec}} = p^2(1-p)^2 + 2p^3(1-p)$$
$$= p^2 - p^4.$$
$$= 0.000\,099\,99 \qquad \text{if } p = 0.01,$$

and so only one word in about 10 000 will be accepted with errors undetected.

   In the early days of coding theory, a popular scheme, when possible, was detection and retransmission. With only a moderately good code, it is possible to run such a scheme for several hours with hardly any undetected errors. The difficulty is that incoming data gets held up by requests for retransmission and this can cause buffer overflows.

   The retransmission probability for an $[n, k]$-code is given by

$$P_{\text{retrans}} = 1 - (1-p)^n - P_{\text{undetec}}.$$

For example, with the [4, 2]-code of Example 6.6, if $p = 0.01$, then $P_{\text{retrans}} \cong 0.04$ and so about 4% of messages have to be retransmitted. This percentage increases for longer codes; e.g. if we used a [24, 12]-code, then P_{retrans} would be over 20%.

   A compromise scheme incorporating both error correction and detection, called 'incomplete decoding', will be described in Chapter 7.

Concluding remark on Chapter 6

The birth of coding theory was inspired by the classic paper of Claude Shannon, of Bell Telephone Laboratories, in 1948. In fact, this single paper gave rise to *two* whole new subjects. The first, *information theory*, is a direct extension of Shannon's work, relying mainly on ideas from probability theory, and this will not be pursued here. The second, *coding theory*, relies mainly on ideas from pure mathematics and, while retaining some links with information theory, has developed largely independently.

Exercises 6

6.1 Construct standard arrays for codes having each of the following generator matrices:

$$G_1 = \begin{bmatrix} 1 & 0 \\ 0 & 1 \end{bmatrix}, \quad G_2 = \begin{bmatrix} 1 & 0 & 1 \\ 0 & 1 & 1 \end{bmatrix}, \quad G_3 = \begin{bmatrix} 1 & 0 & 1 & 1 & 0 \\ 0 & 1 & 0 & 1 & 1 \end{bmatrix}.$$

Using the third array:
(i) decode the received vectors 11111 and 01011,
(ii) give examples of
  (a) two errors occurring in a codeword and being corrected,
  (b) two errors occurring in a codeword and not being corrected.

6.2 If the symbol error probability of a binary symmetric channel is p, calculate the probability, for each of the three codes of Exercise 6.1, that any received vector will be decoded as the codeword which was sent. Evaluate these probabilities for $p = 0.01$.

Now suppose each code is used purely for error detection. Calculate the respective probabilities that the received vector is a codeword different from that sent (and evaluate for $p = 0.01$). Comment on the relative merits of the three codes.

6.3 We have assumed that, for a binary symmetric channel, the symbol error probability p is less than $\frac{1}{2}$. Can an error-correcting code be used to reduce the number of messages received in error if (i) $p = \frac{1}{2}$, (ii) $p > \frac{1}{2}$?

6.4 Suppose C is a binary $[n, k]$-code with minimum distance $2t + 1$ (or $2t + 2$). Given that p is very small, show that an approximate value of $P_{err}(C)$ is

$$\left(\binom{n}{t + 1} - \alpha_{t+1} \right) p^{t+1},$$

where α_{t+1} is the number of coset leaders of C of weight $t+1$.

6.5 Show that if the perfect binary $[7, 4]$-code is used for error detection, then if $p = 0.01$, $P_{undetec} \cong 0.00\,000\,68$ and about 7% of words have to be retransmitted.

   [*Hint:* The codewords of such a code are listed in Example 2.23.]

6.6 We shall see in Chapter 9 that there exists a perfect binary $[23, 12, 7]$-code, called the binary Golay code. Show that, if $p = 0.01$, the word error rate for this code is about $0.000\,08$.

6.7 If standard array decoding is used for a binary $[n, k]$-code and the messages are equally likely, show that P_{symb} does not depend on which codeword was sent and that

$$P_{symb} = \frac{1}{k} \sum_{i=1}^{2^k} F_i P_i,$$

   where F_i is the weight of the first k places of the codeword at the top of the ith column of the standard array, and P_i is the probability that the error vector is in this ith column.

6.8 Show that if $p = 0.01$, the code of Example 6.5 has

$$P_{symb} \cong 0.005\,3.$$

6.9 Show that for a binary $[n, k]$-code,

$$\frac{1}{k} P_{err} \leqslant P_{symb} \leqslant P_{err}.$$

7 The dual code, the parity-check matrix, and syndrome decoding

As well as specifying a linear code by a generator matrix, there is another important way of specifying it—by a parity-check matrix. First we need some definitions.

The *inner product* $\mathbf{u} \cdot \mathbf{v}$ of vectors $\mathbf{u} = u_1 u_2 \cdots u_n$ and $\mathbf{v} = v_1 v_2 \cdots v_n$ in $V(n, q)$ is the scalar (i.e. element of $GF(q)$) defined by

$$\mathbf{u} \cdot \mathbf{v} = u_1 v_1 + u_2 v_2 + \cdots + u_n v_n.$$

For example, in $V(4, 2)$, $(1001) \cdot (1101) = 0$,

$$(1111) \cdot (1110) = 1,$$

and in $V(4, 3)$, $(2011) \cdot (1210) = 0$,

$$(1212) \cdot (2121) = 2.$$

If $\mathbf{u} \cdot \mathbf{v} = 0$, then $\mathbf{u}$ and $\mathbf{v}$ are called *orthogonal*.

The proof of the following lemma is left as a straightforward exercise for the reader.

Lemma 7.1 For any $\mathbf{u}$, $\mathbf{v}$ and $\mathbf{w}$ in $V(n, q)$ and $\lambda, \mu \in GF(q)$,
(i) $\mathbf{u} \cdot \mathbf{v} = \mathbf{v} \cdot \mathbf{u}$
(ii) $(\lambda \mathbf{u} + \mu \mathbf{v}) \cdot \mathbf{w} = \lambda(\mathbf{u} \cdot \mathbf{w}) + \mu(\mathbf{v} \cdot \mathbf{w})$.

Given a linear $[n, k]$-code C, the *dual code* of C, denoted by $C^\perp$, is defined to be the set of those vectors of $V(n, q)$ which are orthogonal to every codeword of C, i.e.

$$C^\perp = \{\mathbf{v} \in V(n, q) \mid \mathbf{v} \cdot \mathbf{u} = 0 \quad \text{for all } \mathbf{u} \in C\}.$$

After a preliminary lemma, we shall show that $C^\perp$ is a linear code of dimension $n - k$.

Lemma 7.2 Suppose C is an $[n, k]$-code having a generator matrix G. Then a vector $\mathbf{v}$ of $V(n, q)$ belongs to $C^\perp$ if and only if $\mathbf{v}$ is orthogonal to every row of G; i.e. $\mathbf{v} \in C^\perp \Leftrightarrow \mathbf{v} G^T = 0$, where G^T denotes the transpose of G.

Proof　The 'only if' part is obvious since the rows of G are codewords. For the 'if' part, suppose that the rows of G are $\mathbf{r}_1, \mathbf{r}_2, \ldots, \mathbf{r}_k$ and that $\mathbf{v} \cdot \mathbf{r}_i = 0$ for each i. If $\mathbf{u}$ is any codeword of C, then $\mathbf{u} = \sum_{i=1}^{k} \lambda_i \mathbf{r}_i$ for some scalars λ_i and so

$$\mathbf{v} \cdot \mathbf{u} = \sum_{i=1}^{k} \lambda_i (\mathbf{v} \cdot \mathbf{r}_i) \qquad \text{(by Lemma 7.1(ii))}$$

$$= \sum_{i=1}^{k} \lambda_i 0 = 0.$$

Hence $\mathbf{v}$ is orthogonal to every codeword of C and so is in $C^\perp$.

Theorem 7.3　Suppose C is an $[n, k]$-code over $GF(q)$. Then the dual code $C^\perp$ of C is a linear $[n, n-k]$-code.

Proof　First we show that $C^\perp$ is a linear code.
　Suppose $\mathbf{v}_1, \mathbf{v}_2 \in C^\perp$ and $\lambda, \mu \in GF(q)$. Then, for all $\mathbf{u} \in C$,

$$(\lambda \mathbf{v}_1 + \mu \mathbf{v}_2) \cdot \mathbf{u} = \lambda(\mathbf{v}_1 \cdot \mathbf{u}) + \mu(\mathbf{v}_2 \cdot \mathbf{u}) \qquad \text{(by Lemma 7.1)}$$

$$= \lambda 0 + \mu 0 = 0.$$

Hence $\lambda \mathbf{v}_1 + \mu \mathbf{v}_2 \in C^\perp$, and so $C^\perp$ is linear, by Exercise 4.1.
　We now show that $C^\perp$ has dimension $n - k$. Let $G = [g_{ij}]$ be a generator matrix for C. Then, by Lemma 7.2, the elements of $C^\perp$ are the vectors $\mathbf{v} = v_1 v_2 \cdots v_n$ satisfying

$$\sum_{j=1}^{n} g_{ij} v_j = 0 \qquad \text{for } i = 1, 2, \ldots, k.$$

This is a system of k independent homogeneous equations in n unknowns and it is a standard result in linear algebra that the solution space $C^\perp$ has dimension $n - k$. For completeness we show this to be so as follows.
　It is clear that if codes C_1 and C_2 are equivalent, then so also are $C_1^\perp$ and $C_2^\perp$. Hence it is enough to show that $\dim(C^\perp) = n - k$ in the case where C has a standard form generator matrix

$$G = \begin{bmatrix} 1 & \cdots & 0 & a_{11} & \cdots & a_{1,n-k} \\ \vdots & \ddots & \vdots & \vdots & & \vdots \\ 0 & \cdots & 1 & a_{k1} & \cdots & a_{k,n-k} \end{bmatrix}$$

Then

$$C^\perp = \Big\{(v_1, v_2, \ldots, v_n) \in V(n, q) \mid v_i$$
$$+ \sum_{j=1}^{n-k} a_{ij}v_{k+j} = 0, \quad i = 1, 2, \ldots, k\Big\}.$$

Clearly for each of the q^{n-k} choices of $(v_{k+1}, \ldots, v_n)$, there is a unique vector $(v_1, v_2, \ldots, v_n)$ in $C^\perp$. Hence $|C^\perp| = q^{n-k}$ and so $\dim(C^\perp) = n - k$.

Examples 7.4 It is easily checked that
 (i) if

$$C = \begin{cases} 0000 \\ 1100 \\ 0011 \\ 1111 \end{cases}, \text{ then } C^\perp = C.$$

 (ii) if

$$C = \begin{cases} 000 \\ 110 \\ 011 \\ 101 \end{cases}, \text{ then } C^\perp = \begin{cases} 000 \\ 111 \end{cases}.$$

Theorem 7.5 For any $[n, k]$-code C, $(C^\perp)^\perp = C$.

Proof Clearly $C \subseteq (C^\perp)^\perp$ since every vector in C is orthogonal to every vector in $C^\perp$. But $\dim((C^\perp)^\perp) = n - (n - k) = k = \dim C$, and so $C = (C^\perp)^\perp$.

Definition A *parity-check matrix* H for an $[n, k]$-code C is a generator matrix of $C^\perp$.
 Thus H is an $(n - k) \times n$ matrix satisfying $GH^\mathrm{T} = \mathbf{0}$, where H^T denotes the transpose of H and $\mathbf{0}$ is an all-zero matrix. It follows from Lemma 7.2 and Theorem 7.5 that if H is a parity-check matrix of C, then

$$C = \{\mathbf{x} \in V(n, q) \mid \mathbf{x}H^T = \mathbf{0}\}.$$

In this way any linear code is completely specified by a parity-check matrix.

In Example 7.4(i),

$$\begin{bmatrix} 1100 \\ 0011 \end{bmatrix}$$

is both a generator matrix and a parity-check matrix, while in (ii), [111] is a parity-check matrix.

The rows of a parity check matrix are *parity checks* on the codewords; they say that certain linear combinations of the co-ordinates of every codeword are zero. A code is completely specified by a parity-check matrix; e.g. if

$$H = \begin{bmatrix} 1100 \\ 0011 \end{bmatrix},$$

then C is the code

$$\{(x_1, x_2, x_3, x_4) \in V(4, 2) \mid x_1 + x_2 = 0, x_3 + x_4 = 0\}.$$

The equations $x_1 + x_2 = 0$ and $x_3 + x_4 = 0$ are called *parity-check equations*.

If $H = [111]$, then C consists of those vectors of $V(3, 2)$ whose coordinates sum to zero (mod 2). More generally, the even weight code E_n of Exercise 5.2 can be defined to be the set of all vectors $x_1 x_2 \cdots x_n$ of $V(n, 2)$ which satisfy the single parity-check equation

$$x_1 + x_2 + \cdots + x_n = 0.$$

The following theorem gives an easy way of constructing a parity-check matrix for a linear code with given generator matrix, or vice versa.

Theorem 7.6 If $G = [I_k \mid A]$ is the standard form generator matrix of an $[n, k]$-code C, then a parity-check matrix for C is $H = [-A^T \mid I_{n-k}]$.

Proof Suppose

$$G = \begin{bmatrix} 1 & & 0 & a_{11} & \cdots & a_{1,n-k} \\ & \ddots & & \vdots & & \vdots \\ 0 & & 1 & a_{k1} & \cdots & a_{k,n-k} \end{bmatrix}.$$

Let

$$H = \begin{bmatrix} -a_{11} & \cdots & -a_{k1} & 1 & & 0 \\ \vdots & & \vdots & & \ddots & \\ -a_{1,n-k} & \cdots & -a_{k,n-k} & 0 & & 1 \end{bmatrix}.$$

Then H has the size required of a parity-check matrix and its rows are linearly independent. Hence it is enough to show that every row of H is orthogonal to every row of G. But the inner product of the ith row of G with the jth row of H is

$$0 + \cdots + 0 + (-a_{ij}) + 0 + \cdots + 0 + a_{ij} + 0 + \cdots + 0 = 0.$$

Example 7.7 The code of Example 5.6(ii) has standard form generator matrix

$$G = \begin{bmatrix} I_4 & \begin{matrix} 101 \\ 111 \\ 110 \\ 011 \end{matrix} \end{bmatrix}.$$

Hence a parity-check matrix is

$$H = \begin{bmatrix} 1110 \\ 0111 \\ 1101 \end{bmatrix} \quad I_3 \end{bmatrix}.$$

(Note that the minus signs are unnecessary in the binary case.)

Definition A parity-check matrix H is said to be in *standard form* if $H = [B \mid I_{n-k}]$.

The proof of Theorem 7.6 shows that if a code is specified by a parity-check matrix in standard form $H = [B \mid I_{n-k}]$, then a generator matrix for the code is $G = [I_k \mid -B^T]$. Many codes, e.g. the Hamming codes (see Chapter 8), are most easily defined by specifying a parity-check matrix or, equivalently, a set of parity-check equations. If a code is given by a parity-check matrix H which is not in standard form, then H can be reduced to standard form in the same way as for a generator matrix.

Syndrome decoding

Suppose H is a parity-check matrix of an $[n, k]$-code C. Then for any vector $\mathbf{y} \in V(n, q)$, the $1 \times (n - k)$ row vector

$$S(\mathbf{y}) = \mathbf{y}H^T$$

is called the *syndrome* of $\mathbf{y}$.

Notes (i) If the rows of H are $\mathbf{h}_1, \mathbf{h}_2, \ldots, \mathbf{h}_{n-k}$, then $S(\mathbf{y}) = (\mathbf{y} \cdot \mathbf{h}_1, \mathbf{y} \cdot \mathbf{h}_2, \ldots, \mathbf{y} \cdot \mathbf{h}_{n-k})$.

(ii) $S(\mathbf{y}) = \mathbf{0} \Leftrightarrow \mathbf{y} \in C$.

(iii) Some authors define the syndrome of $\mathbf{y}$ to be the column vector $H\mathbf{y}^T$ (i.e. the transpose of $S(\mathbf{y})$ as defined above).

Lemma 7.8 Two vectors $\mathbf{u}$ and $\mathbf{v}$ are in the same coset of C if and only if they have the same syndrome.

Proof $\mathbf{u}$ and $\mathbf{v}$ are in the same coset

$$\Leftrightarrow \mathbf{u} + C = \mathbf{v} + C$$
$$\Leftrightarrow \mathbf{u} - \mathbf{v} \in C$$
$$\Leftrightarrow (\mathbf{u} - \mathbf{v})H^T = \mathbf{0}$$
$$\Leftrightarrow \mathbf{u}H^T = \mathbf{v}H^T$$
$$\Leftrightarrow S(\mathbf{u}) = S(\mathbf{v}).$$

Corollary 7.9 There is a one-to-one correspondence between cosets and syndromes.

In standard array decoding, if n is small there is no difficulty in locating the received vector $\mathbf{y}$ in the array. But if n is large, we can save a lot of time by using the syndrome to find out which coset (i.e. which row of the array) contains $\mathbf{y}$. We do this as follows.

Calculate the syndrome $S(\mathbf{e})$ for each coset leader $\mathbf{e}$ and extend the standard array by listing the syndromes as an extra column.

Example 7.10 In Example 6.5,

$$G = \begin{bmatrix} 1011 \\ 0101 \end{bmatrix},$$

and so, by Theorem 7.6, a parity-check matrix is

$$H = \begin{bmatrix} 1010 \\ 1101 \end{bmatrix}.$$

Hence the syndromes of the coset leaders (see Example 6.6) are

$$S(0000) = 00$$
$$S(1000) = 11$$
$$S(0100) = 01$$
$$S(0010) = 10.$$

The standard array becomes:

coset leaders				syndromes
0 0 0 0	1 0 1 1	0 1 0 1	1 1 1 0	0 0
1 0 0 0	0 0 1 1	1 1 0 1	0 1 1 0	1 1
0 1 0 0	1 1 1 1	0 0 0 1	1 0 1 0	0 1
0 0 1 0	1 0 0 1	0 1 1 1	1 1 0 0	1 0.

The decoding algorithm is now: when a vector $\mathbf{y}$ is received, calculate $S(\mathbf{y}) = \mathbf{y}H^T$ and locate $S(\mathbf{y})$ in the 'syndromes' column of the array. Locate $\mathbf{y}$ in the corresponding row and decode as the codeword at the top of the column containing $\mathbf{y}$.

For example, if 1111 is received, $S(1111) = 01$, and so 1111 occurs in the third row of the array.

When programming a computer to do standard array decoding, we need store only two columns (syndromes and coset leaders) in the computer memory. This is called a *syndrome look-up table*.

Example 7.10 (*continued*) The syndrome look-up table for this code is

syndrome $\mathbf{z}$	coset leader $f(\mathbf{z})$
0 0	0 0 0 0
1 1	1 0 0 0
0 1	0 1 0 0
1 0	0 0 1 0

The decoding procedure is:

Step 1 For a received vector $\mathbf{y}$ calculate $S(\mathbf{y}) = \mathbf{y}H^T$.

Step 2 Let $z = S(y)$, and locate z in the first column of the look-up table.

Step 3 Decode y as $y - f(z)$.

For example, if $y = 1111$, then $S(y) = 01$ and we decode as $1111 - 0100 = 1011$.

Incomplete decoding

This is a blend of error correction and detection, the latter being used when 'correction' is likely to give the wrong codeword. More precisely, if $d(C) = 2t + 1$ or $2t + 2$, we adopt the following scheme whereby we guarantee the correction of $\leq t$ errors in any codeword and also detect some cases of more than t errors.

We arrange the cosets of the standard array, as usual, in order of increasing weight of the coset leaders, and divide the array into a *top part* comprising those cosets whose leaders have weights $\leq t$ and a *bottom part* comprising the remaining cosets. If the received vector y is in the top part, we decode it as usual (thus assuming $\leq t$ errors); if y is in the bottom part, we conclude that more than t errors have occurred and ask for re-transmission.

Example 7.11 Let C be the binary code with generator matrix

$$\begin{bmatrix} 10110 \\ 01011 \end{bmatrix}.$$

A standard array for C is

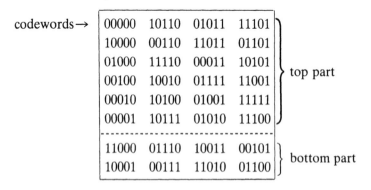

codewords →

00000	10110	01011	11101	top part
10000	00110	11011	01101	
01000	11110	00011	10101	
00100	10010	01111	11001	
00010	10100	01001	11111	
00001	10111	01010	11100	
11000	01110	10011	00101	bottom part
10001	00111	11010	01100	

↑
coset leaders

If 11110 is received, we decode as 10110, but if 10011 is received, we seek re-transmission. Note that in this example, if a received vector **y** found in the bottom part were 'corrected', then owing to the presence of two vectors of weight 2 in each such coset, we would have less than an 'evens' chance of decoding **y** to the codeword sent; e.g. if 10011 is received, then, assuming two errors, the codeword sent could have been 01011 or 10110.

An incomplete decoding scheme is particularly well-suited to a code with even minimum distance. For if $d(C) = 2t + 2$, then it will guarantee to correct up to t errors and simultaneously to detect any $t + 1$ errors.

When we carry out incomplete decoding by means of a syndrome look-up table, we can dispense with the standard array not only in the decoding scheme but also in the actual construction of the table. This is because we know precisely what the coset leaders are in the top part of the array (namely, all those vectors of weight $\leq t$), while those in the bottom half are not used in decoding and so need not be found. In other words we just store the 'top part' of a syndrome look-up table as we now illustrate.

Example 7.11 (continued) By Theorem 7.6, a parity-check matrix is
$$H = \begin{bmatrix} 10100 \\ 11010 \\ 01001 \end{bmatrix}.$$

Calculating syndromes of coset leaders via $S(\mathbf{y}) = \mathbf{y}H^T$, we get (the 'top part' of) the syndrome look-up table thus (the second column was written down first):

syndrome **z**	coset leader $f(\mathbf{z})$
0 0 0	0 0 0 0 0
1 1 0	1 0 0 0 0
0 1 1	0 1 0 0 0
1 0 0	0 0 1 0 0
0 1 0	0 0 0 1 0
0 0 1	0 0 0 0 1

When a vector **y** is received, we calculate $S(\mathbf{y})$ and decode if $S(\mathbf{y})$ appears in the **z** column. If $S(\mathbf{y})$ does not appear, we seek re-transmission. For example, (i) if $\mathbf{y} = 11111$, then $S(\mathbf{y}) = 010$ and so we decode as $11111 - 00010 = 11101$.

(ii)   if $\mathbf{y} = 10011$, then $S(\mathbf{y}) = 101$, which does not appear in the table and so we conclude that at least 2 errors have occurred.

We next consider an interesting non-binary code having a neat syndrome decoding algorithm which does not even require a look-up table. This is the decimal code promised in Example 1.4. Because 10 is not a prime power, the code will be derived from a linear code over $GF(11)$, as was the ISBN code described in Chapter 3, but here the codewords satisfy two parity-check equations instead of just one.

Example 7.12   Consider the linear $[10, 8]$-code over $GF(11)$ defined to have parity-check matrix

$$H = \begin{bmatrix} 1 & 1 & 1 & 1 & 1 & 1 & 1 & 1 & 1 & 1 \\ 1 & 2 & 3 & 4 & 5 & 6 & 7 & 8 & 9 & 10 \end{bmatrix}.$$

H is deliberately chosen not to have standard form here in order to get a nice decoding algorithm later.

Let C be the 10-ary code obtained from this 11-ary code by omitting all those codewords which contain the digit '10'. In other words, C consists of all 10-digit decimal numbers $\mathbf{x} = x_1 x_2 \cdots x_{10}$ satisfying the two parity-check equations

$$\sum_{i=1}^{10} x_i \equiv 0 \pmod{11} \quad \text{and} \quad \sum_{i=1}^{10} i x_i \equiv 0 \pmod{11}.$$

It can be shown, e.g. via the inclusion–exclusion principle, that C contains $82\,644\,629$ codewords, but we omit the proof of this here. The codewords of C can be listed by finding a generator matrix in standard form. To do this we first put H into standard form via elementary row operations.

$$H \xrightarrow[r_1 \to r_1 + r_2]{} \begin{bmatrix} 2 & 3 & 4 & 5 & 6 & 7 & 8 & 9 & 10 & 0 \\ 1 & 2 & 3 & 4 & 5 & 6 & 7 & 8 & 9 & 10 \end{bmatrix}$$

$$\xrightarrow[\substack{r_1 \to (-1)r_1 \\ r_2 \to (-1)r_2}]{} \begin{bmatrix} 9 & 8 & 7 & 6 & 5 & 4 & 3 & 2 & 1 & 0 \\ 10 & 9 & 8 & 7 & 6 & 5 & 4 & 3 & 2 & 1 \end{bmatrix}$$

$$\xrightarrow[r_2 \to r_2 - 2r_1]{} \begin{bmatrix} 9 & 8 & 7 & 6 & 5 & 4 & 3 & 2 & 1 & 0 \\ 3 & 4 & 5 & 6 & 7 & 8 & 9 & 10 & 0 & 1 \end{bmatrix}.$$

Using Theorem 7.6,

$$
G = \begin{bmatrix} & & \begin{array}{cc} 2 & 8 \\ 3 & 7 \\ 4 & 6 \\ 5 & 5 \\ 6 & 4 \\ 7 & 3 \\ 8 & 2 \\ 9 & 1 \end{array} \end{bmatrix}
$$

and so $C = \{(x_1, x_2, \ldots, x_8, \ 2x_1 + 3x_2 + \cdots + 9x_8, \ 8x_1 + 7x_2 + \cdots + x_8)\}$, where $x_1, x_2, \ldots, x_8$ run over the values $0, 1, 2, \ldots, 9$ and those words are omitted which give the digit '10' in either of the last two coordinate places.

We now describe an incomplete syndrome decoding scheme which will correct any single error and which will simultaneously detect any double error arising from the transposition of two digits of a codeword.

Suppose $\mathbf{x} = (x_1, x_2, \ldots, x_{10})$ is the codeword transmitted and $\mathbf{y} = (y_1, y_2, \ldots, y_{10})$ is the received vector. The syndrome

$$
(A, B) = \mathbf{y} H^T = \left(\sum_{i=1}^{10} y_i, \sum_{i=1}^{10} iy_i \right)
$$

is calculated (modulo 11).

Suppose a single error has occurred, so that for some non-zero j and k,

$$
(y_1, y_2, \ldots, y_{10}) = (x_1, \ldots, x_{j-1}, x_j + k, x_{j+1}, \ldots, x_{10}).
$$

Then

$$
A = \sum_{i=1}^{10} y_i = \left(\sum_{i=1}^{10} x_i \right) + k \ \equiv k \pmod{11},
$$

$$
B = \sum_{i=1}^{10} iy_i = \left(\sum_{i=1}^{10} ix_i \right) + jk \equiv jk \pmod{11}.
$$

So the error magnitude k is given by A and the error position j is given by the value of B/A. (The latter is calculated as BA^{-1} as described in Chapter 3). Hence the decoding scheme is, after calculating (A, B) from **y**, as follows:

(1) if $(A, B) = (0, 0)$, then **y** is a codeword and we assume no errors,

(2) if $A \neq 0$ and $B \neq 0$, then we assume a single error which is corrected by subtracting A from the (B/A)th entry of **y**,

(3) if $A = 0$ or $B = 0$ but not both, then we have detected at least two errors.

Case (3) always arises if two digits of a codeword have been transposed, for then $A = 0$ and (as for the ISBN code) $B \neq 0$.

For example, suppose $\mathbf{y} = 0610271355$. We calculate that $A = 8$ and $B = 6$. Hence $B/A = 6 \cdot 8^{-1} = 6 \cdot 7 = 42 = 9$, and so the 9th digit should have been $5 - 8 = -3 = 8$.

Remarks on Example 7.12 (1) Note how much faster is this decoding scheme than the brute-force scheme of comparing the received vector with all codewords. There is no need to store a list of codewords in the memory of the decoder, nor is there even any need to store a syndrome look-up table.

(2) The fact that we are able to correct *any* single error gives an indirect proof that the minimum distance of the code is at least 3. We will see in Example 8.8 that the minimum distance could have been deduced directly by inspection of the parity-check matrix H.

(3) Some further decimal codes will be discussed in Chapter 11.

Exercises 7

7.1 Prove Lemma 7.1.

7.2 Prove that if E_n is the binary even weight code of length n, then $E_n^{\perp}$ is the repetition code of length n.

7.3 Give a very simple scheme for error detection with a linear code, making use of the syndrome.

7.4 For a binary linear code with parity-check matrix H, show that the transpose of the syndrome of a received vector is equal to the sum of those columns of H corresponding to where the errors occurred.

7.5 Construct a syndrome look-up table for the perfect binary [7, 4, 3]-code which has generator matrix

$$\begin{bmatrix} 1 & 0 & 0 & 0 & 1 & 1 & 1 \\ 0 & 1 & 0 & 0 & 1 & 1 & 0 \\ 0 & 0 & 1 & 0 & 1 & 0 & 1 \\ 0 & 0 & 0 & 1 & 0 & 1 & 1 \end{bmatrix}.$$

Use your table to decode the following received vectors:

0000011, 1111111, 1100110, 1010101.

7.6 Let C be the ternary linear code with generator matrix

$$\begin{bmatrix} 1 & 1 & 1 & 0 \\ 2 & 0 & 1 & 1 \end{bmatrix}.$$

(a) Find a generator matrix for C in standard form.
(b) Find a parity-check matrix for C in standard form.
(c) Use syndrome decoding to decode the received vectors 2121, 1201 and 2222.

7.7 Using the code of Example 7.12, decode the received vector 0617960587.

7.8 Example 7.12 shows that $A_{10}(10, 3) \geqslant 82\,644\,629$. Prove that $A_{10}(10, 3) \leqslant 10^8$. Prove also that $A_{11}(10, 3) = 11^8$.

7.9 Show that the decimal code

$$\left\{ (x_1, x_2, \ldots, x_{10}) \in (F_{10})^{10} \,\middle|\, \sum_{i=1}^{10} x_i \right.$$
$$\left. \equiv 0 \,(\mathrm{mod}\ 10), \sum_{i=1}^{10} ix_i \equiv 0 \,(\mathrm{mod}\ 10) \right\}.$$

is *not* a single-error-correcting code.

7.10 Suppose a certain binary channel accepts words of length 7 and that the only kind of error vector ever observed is one of the eight vectors 0000000, 0000001, 0000011, 0000111, 0001111, 0011111, 0111111, 1111111. Design a binary linear [7, k]-code which will correct all such errors with as large a rate as possible.

7.11 Suppose C is a binary code with parity-check matrix H. Show that the *extended code* $\hat{C}$, obtained from C by adding an overall parity-check, has parity-check matrix

$$\bar{H} = \left[\begin{array}{ccc|c} & & & 0 \\ & H & & 0 \\ & & & \vdots \\ & & & 0 \\ \hline 1 & 1 \cdots 1 & & 1 \end{array} \right].$$

8 The Hamming codes

The Hamming codes are an important family of single-error-correcting codes which are easy to encode and decode. They are linear codes and can be defined over any finite field $GF(q)$ but, for simplicity, we first restrict our attention to the binary case.

A Hamming code is most conveniently defined by specifying its parity-check matrix:

Definition Let r be a positive integer and let H be an $r \times (2^r - 1)$ matrix whose columns are the distinct non-zero vectors of $V(r, 2)$. Then the code having H as its parity-check matrix is called a *binary Hamming code* and is denoted by Ham $(r, 2)$.

(We shall later generalize this to define Ham (r, q) for any prime power q.)

Notes (i) Ham $(r, 2)$ has length $n = 2^r - 1$ and dimension $k = n - r$. Thus $r = n - k$ is the number of check symbols in each codeword and is also known as the *redundancy* of the code.

(ii) Since the columns of H may be taken in any order, the code Ham $(r, 2)$ is, for given redundancy r, any one of a number of equivalent codes.

Examples 8.1 (i) $r = 2$: $H = \begin{bmatrix} 1 & 1 & 0 \\ 1 & 0 & 1 \end{bmatrix}$.

By Theorem 7.6, $G = [111]$, and so Ham $(2, 2)$ is just the binary triple repetition code.

(ii) $r = 3$: a parity-check matrix for Ham $(3, 2)$ is

$$H = \begin{bmatrix} 0 & 0 & 0 & 1 & 1 & 1 & 1 \\ 0 & 1 & 1 & 0 & 0 & 1 & 1 \\ 1 & 0 & 1 & 0 & 1 & 0 & 1 \end{bmatrix}.$$

Here we have taken the columns in the natural order of increasing binary numbers (from 1 to 7). To get H in standard form we take the columns in a different order:

$$H = \begin{bmatrix} 0 & 1 & 1 & 1 & 1 & 0 & 0 \\ 1 & 0 & 1 & 1 & 0 & 1 & 0 \\ 1 & 1 & 0 & 1 & 0 & 0 & 1 \end{bmatrix}.$$

Hence, by Theorem 7.6, a generator matrix for Ham $(3, 2)$ is

$$G = \begin{bmatrix} 1 & 0 & 0 & 0 & 0 & 1 & 1 \\ 0 & 1 & 0 & 0 & 1 & 0 & 1 \\ 0 & 0 & 1 & 0 & 1 & 1 & 0 \\ 0 & 0 & 0 & 1 & 1 & 1 & 1 \end{bmatrix}.$$

It is easily seen that Ham $(3, 2)$ is equivalent to the perfect $[7, 4, 3]$-code of Example 5.6 (and Examples 2.23 and 5.3). We show next that *all* the binary Hamming codes are perfect.

Theorem 8.2   The binary Hamming code Ham $(r, 2)$, for $r \geq 2$,

(i)   is a $[2^r - 1, 2^r - 1 - r]$-code;
(ii) has minimum distance 3 (hence is single-error-correcting);
(iii) is a perfect code.

Proof   (i)   By definition, Ham $(r, 2)^\perp$ is a $[2^r - 1, r]$-code and so Ham $(r, 2)$ is a $[2^r - 1, 2^r - 1 - r]$-code.

(ii)   Since Ham $(r, 2)$ is a linear code, it is enough, by Theorem 5.2, to show that every non-zero codeword has weight ≥ 3. We do this by showing that Ham $(r, 2)$ has no codewords of weight 1 or 2.

Suppose Ham $(r, 2)$ has a codeword $\mathbf{x}$ of weight 1, say

$$\mathbf{x} = 00 \cdots 010 \cdots 0 \quad \text{(with 1 in the } i\text{th place)}.$$

Since $\mathbf{x}$ is orthogonal to every row of the parity-check matrix H, the ith entry of every row of H is zero. Hence the ith column of H is the all-zero vector, contradicting the definition of H.

Now suppose Ham $(r, 2)$ has a codeword $\mathbf{x}$ of weight 2, say

$$\mathbf{x} = 0 \cdots 010 \cdots 010 \cdots 0 \quad \text{(with 1s in the } i\text{th and } j\text{th places)}.$$

Denoting the sth row of H by $[h_{s1} h_{s2} \cdots h_{sn}]$, we have, since $\mathbf{x}$ is orthogonal to each such row,

that is
$$h_{si} + h_{sj} = 0 \pmod{2} \qquad \text{for } s = 1, 2, \ldots, r;$$
$$h_{si} = h_{sj} \pmod{2} \qquad \text{for } s = 1, 2, \ldots, r.$$

Hence the ith and jth columns of H are identical, again contradicting the definition of H.

Thus $d(\text{Ham}\,(r, 2)) \geqslant 3$. On the other hand, Ham $(r, 2)$ does contain codewords of weight 3. For example, if the first three columns of H are

$$
\begin{array}{ccc}
0 & 0 & 0 \\
\vdots & \vdots & \vdots \\
0 & 0 & 0 \\
0 & 1 & 1 \\
1 & 0 & 1
\end{array}
$$

then the vector $11100 \cdots 0$ is orthogonal to every row of H and so belongs to Ham $(r, 2)$.

(iii) To show Ham $(r, 2)$ is perfect, it is enough to show that equality holds in the sphere-packing bound (2.18). With $t = 1$, $n = 2^r - 1$ and $M = 2^{n-r}$, the left-hand side of (2.18) becomes

$$2^{n-r}\left(1 + \binom{n}{1}\right) = 2^{n-r}(1 + n) = 2^{n-r}(1 + 2^r - 1) = 2^n,$$

which is equal to the right-hand side of (2.18).

Decoding with a binary Hamming code

Since Ham $(r, 2)$ is a perfect single-error-correcting code, the coset leaders are precisely the $2^r(=n + 1)$ vectors of $V(n, 2)$ of weight $\leqslant 1$.

The syndrome of the vector $0 \cdots 010 \cdots 0$ (with 1 in the jth place) is $(0 \cdots 010 \cdots 0)H^T$, which is just the transpose of the jth column of H.

Hence, if the columns of H are arranged in order of increasing binary numbers (i.e. the jth column of H is just the binary representation of j), then we have the following nice decoding algorithm.

Step 1 When a vector $\mathbf{y}$ is received, calculate its syndrome $S(\mathbf{y}) = \mathbf{y}H^T$.

Step 2   If $S(\mathbf{y}) = \mathbf{0}$, then assume $\mathbf{y}$ was the codeword sent.
Step 3   If $S(\mathbf{y}) \neq \mathbf{0}$, then, assuming a single error, $S(\mathbf{y})$ gives the binary representation of the error position and so the error can be corrected.

$$\text{For example, with } H = \begin{bmatrix} 0 & 0 & 0 & 1 & 1 & 1 & 1 \\ 0 & 1 & 1 & 0 & 0 & 1 & 1 \\ 1 & 0 & 1 & 0 & 1 & 0 & 1 \end{bmatrix},$$

if $\mathbf{y} = 1101011$, then $S(\mathbf{y}) = 110$, indicating an error in the sixth position and so we decode $\mathbf{y}$ as 1101001.

Extended binary Hamming codes

The *extended binary Hamming code* Hâm $(r, 2)$ is the code obtained from Ham $(r, 2)$ by adding an overall parity-check.

As in the proof of Theorem 2.7, the minimum distance is increased from 3 to 4. Also, by Exercise 5.4, the extended code is linear and so Hâm $(r, 2)$ is a $[2^r, 2^r - 1 - r, 4]$-code.

We shall see in Exercise 8.4 that the extended code Hâm $(r, 2)$ is no better than Ham $(r, 2)$ when used for complete decoding. In fact, it is inferior since an extra check digit is required for each codeword, thus slowing down the rate of transmission of information. However, having minimum distance 4, Hâm $(r, 2)$ is ideally suited for incomplete decoding, as described in Chapter 7, for it can simultaneously correct any single error and detect any double error.

Let H be a parity-check matrix for Ham $(r, 2)$. By Exercise 7.11, a parity-check matrix $\bar{H}$ for the extended code may be obtained from H via

$$H \rightarrow \bar{H} = \begin{bmatrix} & & 0 \\ & & 0 \\ & H & \vdots \\ & & 0 \\ 1 1 & \cdots & 1 1 \end{bmatrix}.$$

The last row gives the overall parity-check equation on codewords, i.e. $x_1 + x_2 + \cdots + x_{n+1} = 0$.

If H is taken with columns in increasing order of binary

numbers, there is a neat incomplete decoding algorithm, illustrated for $r = 3$ below, to correct any single error and at the same time detect any double error.

Example 8.3 Hâm $(3, 2)$ has the parity-check matrix

$$\bar{H} = \begin{bmatrix} 0 & 0 & 0 & 1 & 1 & 1 & 1 & 0 \\ 0 & 1 & 1 & 0 & 0 & 1 & 1 & 0 \\ 1 & 0 & 1 & 0 & 1 & 0 & 1 & 0 \\ 1 & 1 & 1 & 1 & 1 & 1 & 1 & 1 \end{bmatrix}.$$

The syndrome of the error vector $0\,0\cdots0\,1\,0\cdots0$ (with 1 in the jth place) is just the transpose of the jth column of $\bar{H}$. The incomplete decoding algorithm is as follows. Suppose the received vector is $\mathbf{y}$. The syndrome $S(\mathbf{y}) = \mathbf{y}\bar{H}^T$ is calculated. Suppose $S(\mathbf{y}) = (s_1, s_2, s_3, s_4)$. Then

   (i) if $s_4 = 0$ and $(s_1, s_2, s_3) = \mathbf{0}$, assume no errors,

   (ii) if $s_4 = 0$ and $(s_1, s_2, s_3) \neq \mathbf{0}$, assume at least two errors have occurred and seek retransmission,

   (iii) if $s_4 = 1$ and $(s_1, s_2, s_3) = \mathbf{0}$, assume a single error in the last place,

   (iv) if $s_4 = 1$ and $(s_1, s_2, s_3) \neq \mathbf{0}$, assume a single error in the jth place, where j is the number whose binary representation is (s_1, s_2, s_3).

A fundamental theorem

Before defining Hamming codes over an arbitrary field $GF(q)$, we establish a fundamental relationship between the minimum distance of a linear code and a linear independence property of the columns of a parity-check matrix. This result will also be important in later chapters.

Theorem 8.4 Suppose C is a linear $[n, k]$-code over $GF(q)$ with parity-check matrix H. Then the minimum distance of C is d if and only if any $d - 1$ columns of H are linearly independent but some d columns are linearly dependent.

Proof By Theorem 5.2, the minimum distance of C is equal to

the smallest of the weights of the non-zero codewords. Let
$\mathbf{x} = x_1 x_2 \cdots x_n$ be a vector in $V(n, q)$. Then

$$\mathbf{x} \in C \Leftrightarrow \mathbf{x} H^T = \mathbf{0}$$
$$\Leftrightarrow x_1 \mathbf{H}_1 + x_2 \mathbf{H}_2 + \cdots + x_n \mathbf{H}_n = \mathbf{0},$$

where $\mathbf{H}_1, \mathbf{H}_2, \ldots, \mathbf{H}_n$ denote the columns of H.

Thus, corresponding to each codeword $\mathbf{x}$ of weight d, there is a
set of d linearly dependent columns of H. On the other hand if
there existed a set of $d - 1$ linearly dependent columns of H, say
$\mathbf{H}_{i_1}, \mathbf{H}_{i_2}, \ldots, \mathbf{H}_{i_{d-1}}$, then there would exist scalars $x_{i_1}, x_{i_2}, \ldots,$
$x_{i_{d-1}}$, not all zero, such that

$$x_{i_1} \mathbf{H}_{i_1} + x_{i_2} \mathbf{H}_{i_2} + \cdots + x_{i_{d-1}} \mathbf{H}_{i_{d-1}} = \mathbf{0}.$$

But then the vector $\mathbf{x} = (0 \cdots 0 x_{i_1} 0 \cdots 0 x_{i_2} 0 \cdots 0 x_{i_{d-1}} 0 \cdots 0)$,
having x_{i_j} in the i_jth position for $j = 1, 2, \ldots, d - 1$, and 0s
elsewhere, would satisfy $\mathbf{x} H^T = \mathbf{0}$ and so would be a non-zero
codeword of weight less than d.

Theorem 8.4 not only provides a means of establishing the
minimum distance of a specific linear code when H is given, but
also provides a means of constructing the parity-check matrix to
provide a code of guaranteed minimum distance. We concentrate
here on the case $d = 3$, leaving a discussion of the general case
until Chapter 14.

q-ary Hamming codes

In order that C be a linear code with minimum distance 3, we
require that any two columns of a parity-check matrix H be
linearly independent. Thus the columns of H must be non-zero
and no column must be a scalar multiple of another (cf. Exercise
4.4). For fixed redundancy r, let us try to construct an $[n, n -
r, 3]$-code over $GF(q)$ with n as large as possible by finding as
large a set as possible of non-zero vectors of $V(r, q)$ such that
none is a scalar multiple of another.

Any non-zero vector $\mathbf{v}$ in $V(r, q)$ has exactly $q - 1$ non-zero
scalar multiples, forming the set $\{\lambda \mathbf{v} \mid \lambda \in GF(q), \lambda \neq 0\}$. In fact,
the $q^r - 1$ non-zero vectors of $V(r, q)$ may be partitioned into
$(q^r - 1)/(q - 1)$ such sets, which we will call classes, such that
two vectors are scalar multiples of each other if and only if they

are in the same class. For example, in $V(2, 3)$, with vectors written as columns, these classes are

$$\left\{ \binom{0}{1}, \binom{0}{2} \right\}, \quad \left\{ \binom{1}{0}, \binom{2}{0} \right\}, \quad \left\{ \binom{1}{1}, \binom{2}{2} \right\} \quad \text{and} \quad \left\{ \binom{1}{2}, \binom{2}{1} \right\}.$$

By choosing one vector from each class we obtain a set of $(q^r - 1)/(q - 1)$ vectors, no two of which are linearly dependent. Hence, by Theorem 8.4, taking these as the columns of H gives a parity-check matrix for a $[(q^r - 1)/(q - 1), (q^r - 1)/(q - 1) - r, 3]$-code. This code is called a q-ary *Hamming code* and is denoted by Ham (r, q).

Note that different parity-check matrices may be chosen to define Ham (r, q) for given r and q, but any such matrix may clearly be obtained from another one by means of a permutation of the columns and/or the multiplication of some columns by non-zero scalars. Thus the Hamming codes are linear codes which are uniquely defined, up to equivalence, by their parameters.

An easy way to write down a parity-check matrix for Ham (r, q) is to list as columns (e.g. in lexicographical order) all non-zero r-tuples in $V(r, q)$ with first non-zero entry equal to 1. This must work because within each class of $q - 1$ scalar multiples there is exactly one vector having 1 as its first non-zero entry.

Examples 8.5 (i) A parity-check matrix for Ham $(2, 3)$ is

$$\begin{bmatrix} 0 & 1 & 1 & 1 \\ 1 & 0 & 1 & 2 \end{bmatrix}.$$

(ii) A parity check matrix for Ham $(2, 11)$ is

$$\begin{bmatrix} 0 & 1 & 1 & 1 & 1 & 1 & 1 & 1 & 1 & 1 & 1 & 1 \\ 1 & 0 & 1 & 2 & 3 & 4 & 5 & 6 & 7 & 8 & 9 & 10 \end{bmatrix}.$$

(iii) A parity-check matrix for Ham $(3, 3)$ is

$$\begin{bmatrix} 0 & 0 & 0 & 0 & 1 & 1 & 1 & 1 & 1 & 1 & 1 & 1 & 1 \\ 0 & 1 & 1 & 1 & 0 & 0 & 0 & 1 & 1 & 1 & 2 & 2 & 2 \\ 1 & 0 & 1 & 2 & 0 & 1 & 2 & 0 & 1 & 2 & 0 & 1 & 2 \end{bmatrix}.$$

Theorem 8.6 Ham (r, q) is a perfect single-error-correcting code.

Proof Ham (r, q) was constructed to be an $(n, M, 3)$-code with $n = (q^r - 1)/(q - 1)$ and $M = q^{n-r}$. With $t = 1$, the left-hand side of the sphere-packing bound (2.17) becomes

$$q^{n-r}(1 + n(q - 1)) = q^{n-r}(1 + q^r - 1)$$
$$= q^n,$$

which is the right-hand side of (2.17), and so Ham (r, q) is a perfect code.

Corollary 8.7 If q is a prime power and if $n = (q^r - 1)/(q - 1)$, for some integer $r \geq 2$, then

$$A_q(n, 3) = q^{n-r}.$$

Thus, if q is a prime power and $d = 3$, then the main coding theory problem, that of finding $A_q(n, 3)$, is solved for an infinite sequence of values of n. In particular, we have now established a further entry of Table 2.4, namely $A_2(15, 3) = 2^{11} = 2048$.

Decoding with a q-ary Hamming code

Since a Hamming code is a perfect single-error correcting code, the coset leaders, other than $\mathbf{0}$, are precisely the vectors of weight 1. The syndrome of such a coset leader is

$$S(0 \cdots 0b0 \cdots 0) = (0 \cdots 0b0 \cdots 0)H^T = b\mathbf{H}_j^T,$$
$$\uparrow$$
$$j\text{th place}$$

where $\mathbf{H}_j$ denotes the jth column of H.

So the decoding scheme is as follows. Given a received vector $\mathbf{y}$, calculate $S(\mathbf{y}) = \mathbf{y}H^T$. If $S(\mathbf{y}) = \mathbf{0}$, assume no errors. If $S(\mathbf{y}) \neq \mathbf{0}$, then $S(\mathbf{y}) = b\mathbf{H}_j^T$ for some b and j and the assumed single error is corrected by subtracting b from the jth entry of $\mathbf{y}$.

For example, suppose $q = 5$ and

$$H = \begin{bmatrix} 0 & 1 & 1 & 1 & 1 & 1 \\ 1 & 0 & 1 & 2 & 3 & 4 \end{bmatrix}.$$

Suppose the received vector is $\mathbf{y} = 203031$. Then $S(\mathbf{y}) = (2, 3) = 2(1, 4)$ and so we decode $\mathbf{y}$ as 203034.

Shortening a code

Shortening a code can be a useful device if we desire a code of given length and minimum distance and if we know of a good code with greater length and the same minimum distance.

Suppose C is a q-ary (n, M, d)-code. Consider a fixed coordinate position, the jth say, and a fixed symbol λ of the alphabet. Then, if we take all the codewords of C having λ in the jth position and then delete this jth coordinate from these codewords, we will get a code C' of length $n - 1$ with, in general, fewer codewords but the same minimum distance. C' is called a *shortened code* of C.

If C is a linear $[n, k, d]$-code, and if the deleted symbol is 0, then the shortened code C' will also be linear; C' will be an $[n - 1, k - 1, d']$-code, where d' will in general be the same as d (it may occasionally be greater than d). If C has parity-check matrix H, then it is easy to see that a parity-check matrix of C' is obtained simply by deleting the corresponding column of H.

Example 8.8 Let us have another look at the $[10, 8]$-code over $GF(11)$ considered in Example 7.12. This was defined to have parity-check matrix

$$H = \begin{bmatrix} 1 & 1 & 1 & 1 & 1 & 1 & 1 & 1 & 1 & 1 \\ 1 & 2 & 3 & 4 & 5 & 6 & 7 & 8 & 9 & 10 \end{bmatrix}$$

and it now follows instantly from Theorem 8.4 that this code has minimum distance at least 3, for clearly any two columns of H are linearly independent. In fact, we see that it is a doubly shortened Hamming code, for H is obtained from the parity-check matrix of Ham $(2, 11)$, as given in Example 8.5(ii), by deleting the first two columns. This doubly shortened Hamming code has two practical advantages over Ham $(2, 11)$; first, it has an even simpler decoding algorithm, as described in Example 7.12, and, secondly, it not only corrects any single error but also detects any double error created by the transposition of two digits. On the other hand, Ham $(2, 11)$ has far more codewords than its doubly shortened version.

The 11-ary [10, 8, 3]-code of Example 8.8 is optimal in that the number of its codewords is equal to the value of $A_{11}(10, 3)$ (see Exercise 7.8), a result which is generalized in Exercise 8.10. While shortening an optimal code will certainly not in general produce an optimal code, it is interesting to note a recent result of Best and Brouwer (1977) that the triply shortened binary Hamming code is optimal; thus

$$A_2(2^r - s, 3) = 2^{2^r - r - s} \qquad \text{for } s = 1, 2, 3, 4. \qquad (8.9)$$

For $s = 1$, (8.9) merely states the optimality of Ham $(r, 2)$, of which we are already aware, while for $s = 2$, 3 and 4, (8.9) tells us that three successive shortenings of Ham $(r, 2)$ are also optimal. The result was proved by the use of linear programming, a technique which has been used to great effect recently in obtaining improved upper bounds on $A_2(n, d)$ for a number of cases. For a good introduction to the method, see Chapter 17 of MacWilliams and Sloane (1977).

Taking $r = 4$ in (8.9) gives the values of $A_2(14, 3)$, $A_2(13, 3)$ and $A_2(12, 3)$ as shown in Table 2.4. However, if Ham $(4, 2)$ is shortened four times, the resulting $(11, 128, 3)$-code is not optimal, for we see from Table 2.4 that there exists a binary $(11, 144, 3)$-code.

Concluding remarks on Chapter 8

(1)   Hamming codes were discovered by Hamming (1950) and Golay (1949).

(2)   For simplicity, we began this chapter by introducing only the binary Hamming codes. In a sense some of that material was made redundant by the treatment of q-ary Hamming codes, which included the case $q = 2$; for example, Theorem 8.2 is just a particular case of Theorem 8.6. However, the discussion of the extended Hamming code is applicable only to the binary case, for we cannot in general add an overall parity-check to a q-ary code in such a way as to guarantee an increase in the minimum distance. This is because Lemma 2.6 and hence Theorem 2.7 do not have suitable analogues for non-binary codes.

(3)   By Theorem 8.4, we can construct the parity-check matrix of a q-ary linear code of redundancy r and minimum

distance d by finding a set of (column) vectors of $V(r, q)$ such that any $d - 1$ of them are linearly independent. As we have seen, it is easy to write down such a set of N vectors for $d = 3$ of any size N we wish up to a maximum value of $(q^r - 1)/(q - 1)$.

For $d \geq 4$ also, it is easy enough to construct a set of vectors of $V(r, q)$, any $d - 1$ of which are linearly independent, simply by writing down vectors of $V(r, q)$, one at a time, each time making sure that the new vector is not a linear combination of any $d - 2$ earlier ones. However, this approach is a little naïve for $d \geq 4$, for we are likely to run out of choices for the new vector at a relatively early stage. In fact, the problem of finding the *maximum* possible number of vectors in $V(r, q)$ such that any $d - 1$ are linearly independent is extremely difficult for $d \geq 4$ and very little is known except for cases $r \leq 4$. The problem is of much interest in other branches of mathematics, namely in finite geometries and in the theory of factorial designs in statistics. We shall return to it in Chapter 14.

We can at least use the above-mentioned naïve approach to get a lower bound on the maximum size of a code for given length and minimum distance. This is the Gilbert bound (also called the Gilbert–Varshamov bound), discovered independently by Gilbert (1952) and Varshamov (1957).

Theorem 8.10 Suppose q is a prime power. Then there exists a q-ary $[n, k]$-code with minimum distance at least d provided the following inequality holds:

$$\sum_{i=0}^{d-2} (q - 1)^i \binom{n-1}{i} < q^{n-k}. \qquad (8.11)$$

Proof Suppose q, n, k and d satisfy (8.11). We shall construct an $(n - k) \times n$ matrix H over $GF(q)$ with the property that no $d - 1$ columns are linearly dependent. By Theorem 8.4, this will establish the theorem. Put $r = n - k$. Choose the first column of H to be any non-zero r-tuple in $V(r, q)$. Then choose the second column to be any non-zero r-tuple which is not a scalar multiple of the first. Continue choosing successive columns so that each new column is not a linear combination of any $d - 2$ or fewer previous columns. There are $q - 1$ possible non-zero coefficients and so when we come to try to choose the ith column, those r-tuples not available to us will be the

$$N(i) = 1 + \binom{i-1}{1}(q-1) + \binom{i-1}{2}(q-1)^2$$

$$+ \cdots + \binom{i-1}{d-2}(q-1)^{d-2}$$

linear combinations of $d-2$ or fewer columns from the $i-1$ columns already chosen. Not all of these linear combinations need be distinct vectors, but even in the worst case, where they are distinct, provided $N(i)$ is less than the total number q^r of all r-tuples, then an ith column can be added to the matrix. Thus, since (8.11) holds, we will reach a matrix H having n columns, as required.

The following is an immediate consequence of Theorem 8.10.

Corollary 8.12 If q is a prime-power, then

$$A_q(n, d) \geq q^{k_1},$$

where k_1 is the largest integer k satisfying

$$q^k < q^n \bigg/ \bigg(\sum_{i=0}^{d-2} (q-1)^i \binom{n-1}{i} \bigg).$$

Corollary 8.12 gives a general lower bound on $A_q(n, d)$ when q is a prime-power and is the best available for large n (see, e.g., Chapter 17, Theorem 30 of MacWilliams and Sloane 1977). However, for specific values of q, n and d one can usually do much better by constructing a good code in some other way. For example, taking $q = 2$, $n = 13$ and $d = 5$, Corollary 8.12 promises only the existence of a binary $(13, M, 5)$-code with $M = 16$, whereas we see from Table 2.4 that the actual value of $A_2(13, 5)$ is 64. We shall construct such an optimal binary $(13, 64, 5)$-code in Exercise 9.10.

For a weaker version of the Gilbert–Varshamov bound, but one which applies for any size q of alphabet, see Exercise 8.12.

Exercises 8

8.1 Write down a parity-check matrix for the binary [15, 11]-Hamming code. Explain how the code can be used to correct any single error in a codeword. What happens if two or more errors occur in any codeword?

8.2 With the code of Example 8.3, use an incomplete decoding algorithm to decode the following received vectors.

$$11100000, \quad 01110000, \quad 11000000, \quad 00110011.$$

8.3 Show that the code of Examples 2.23, 5.3(ii) and 5.6(ii) is a Hamming code.

8.4 Suppose C is a binary Hamming code of length n and that $\hat{C}$ is its extended code of length $n + 1$. For a binary symmetric channel with symbol error probability p, find $P_{\text{corr}}(C)$ and $P_{\text{corr}}(\hat{C})$ in terms of p and n, and show that, surprisingly, $P_{\text{corr}}(\hat{C}) = P_{\text{corr}}(C)$.

8.5 (i) Write down a parity-check matrix for the 7-ary [8, 6]-Hamming code and use it to decode the received vectors 35234106 and 10521360.

(ii) Write down a parity-check matrix for the 5-ary [31, 28]-Hamming code.

8.6 Use Theorem 8.4 to determine the minimum distance of the binary code with generator matrix

$$\left[\begin{array}{c|cccc} & 1 & 1 & 0 & 0 \\ & 1 & 0 & 1 & 0 \\ & 0 & 1 & 1 & 0 \\ I_7 & 1 & 1 & 1 & 1 \\ & 1 & 1 & 0 & 1 \\ & 0 & 1 & 0 & 1 \\ & 1 & 0 & 0 & 1 \end{array} \right].$$

8.7 Let C_1 be the code over $GF(5)$ generated by

$$\begin{bmatrix} 1 & 2 & 4 & 0 & 3 \\ 0 & 2 & 1 & 4 & 1 \\ 2 & 0 & 3 & 1 & 4 \end{bmatrix}.$$

Let C_2 be the code over $GF(3)$ generated by

$$\begin{bmatrix} 1 & 2 & 0 & 2 & 1 & 0 \\ 2 & 0 & 1 & 2 & 0 & 1 \\ 1 & 1 & 1 & 2 & 1 & 2 \end{bmatrix}.$$

Find a parity-check matrix for each code and determine the minimum distance of each code.

8.8   Use Theorem 8.4 to construct a $[6, 3, 4]$-code over $GF(5)$.

8.9   Let R_r denote the rate of the binary Hamming code Ham $(r, 2)$. Determine $\lim_{r \to \infty} R_r$.

8.10 Prove that if q is a prime power and if $3 \le n \le q + 1$, then

$$A_q(n, 3) = q^{n-2}.$$

8.11 (The 'football pool problem') Suppose there are t football matches and that a *bet* consists of forecasting the outcome, home win (1), away win (2) or draw (X), of each of the t matches. Thus a bet can be regarded as a ternary t-tuple over the alphabet $\{1, 2, X\}$.

The 't-match football pool problem' is the following. 'What is the least number $f(t)$ of bets required to guarantee at least a second prize (i.e. a bet having at most one incorrect forecast)?'

(a)   (i)   By using Hamming codes over $GF(3)$, find the value of $f(t)$ for values of t of the form $(3^r - 1)/2$ for some integer $r \ge 2$; i.e. for $t = 4$, 13, 40, 121, ...

       (ii) Enter in the coupon below a minimum number of bets which will guarantee at least 3 correct forecasts in some bet.

Arsenal	Luton													
Coventry	Ipswich													
Liverpool	Chelsea													
Watford	Everton													

(b)   Show that $23 \le f(5) \le 27$.

       [*Remark:* It was shown by Kamps and Van Lint (1967) that $f(5) = 27$, the proof taking ten pages. The value of $f(t)$ is unknown for $t > 5$ except for values 13, 40, 121, etc., covered by part (a). For some recent work on the bounds for $f(6)$, $f(7)$ and $f(8)$ see Fernandes and Rechtschaffen (1983), Weber (1983), and Blokhuis and Lam (1984).]

8.12 (A weaker, but more general, version of the Gilbert–Varshamov bound). Prove that, for any integer $q \geq 2$,

$$A_q(n, d) \geq q^n \Big/ \left(\sum_{i=0}^{d-1} (q-1)^i \binom{n}{i} \right).$$

[*Remark:* When q is a prime power, this bound is much inferior to that of Corollary 8.12. For example, it guarantees the existence of a binary $(13, M, 5)$-code having only $M = 8$, compared with $M = 16$ given by Corollary 8.12 and a largest possible value of M of 64.]

9 Perfect codes

We recall from Chapter 2 that a q-ary t-error-correcting code of length n is called *perfect* if the spheres of radius t about codewords fill the space $(F_q)^n$ with no overlap; thus a q-ary $(n, M, 2t + 1)$-code is perfect if and only if the *sphere-packing condition*

$$M\left\{1 + (q - 1)n + (q - 1)^2\binom{n}{2} + \cdots + (q - 1)^t\binom{n}{t}\right\} = q^n \quad (9.1)$$

is satisfied.

Apart from being the best codes for their n and d, perfect codes are of much interest to mathematicians, largely because of their associated designs and automorphism groups.

The problem of finding all perfect codes was begun by M. Golay in 1949 but not completed until 1973 (and then only in the case of prime-power alphabets) by J. H. van Lint and A. Tietäväinen. Before giving their final result (Theorem 9.5) we review the perfect codes we already know of and describe two new ones.

The *trivial* perfect codes were defined in Chapter 2 to be binary repetition codes of odd length, codes consisting of a single codeword, or the whole of $(F_q)^n$.

In Chapter 8 we defined the perfect q-ary *Hamming codes* with parameters

$$(n, M, d) = ((q^r - 1)/(q - 1), q^{n-r}, 3),$$

for any integer $r \geq 2$ and any prime power q.

Note that the Hamming parameters satisfy (9.1) for any positive integer q and, while it is conjectured that there do not exist any codes having these parameters for q not a prime-power, this is known to be the case only for $q = 6$ and $r = 2$ (see Theorem 9.12).

A natural approach in looking for further perfect codes was first to seek solutions of (9.1) in integers q, M, n and t; i.e. to find q, n and t such that $\sum_{i=0}^{t} (q - 1)^i\binom{n}{i}$ is a power of q. A limited search by Golay (1949) produced only three feasible sets of parameters (n, M, d) other than the above-mentioned. These were $(23, 2^{12}, 7)$ and $(90, 2^{78}, 5)$ with $q = 2$ and $(11, 3^6, 5)$ with $q = 3$.

[*Remark:* A computer search carried out by van Lint in 1967 showed that these are the only further solutions of the sphere-packing condition with $n \leq 1000$, $t \leq 1000$ and $q \leq 100$.]

In his 1949 paper, Golay was concerned only with linear codes. He exhibited generator matrices, which he presumably had found by trial and error, for codes having the parameters $(23, 2^{12}, 7)$ and $(11, 3^6, 5)$, and he also showed that a linear $[90, 78, 5]$-code over $GF(2)$ could not exist. Remarkably, he did all this, together with generalizing the Hamming codes from those over $GF(2)$ to those over any prime field, in less than one page!

Before describing the two perfect Golay codes, we give a proof, based on that of Golay, of the non-existence of a linear code having the third feasible set of parameters.

Theorem 9.2   There does not exist a binary linear $[90, 78, 5]$-code.

Proof   Suppose H were a parity-check matrix for a binary $[90, 78, 5]$-code. Then H is a 12×90 matrix, whose columns we denote by $\mathbf{H}_1, \mathbf{H}_2, \ldots, \mathbf{H}_{90}$. By Theorem 8.4, any four columns of H are linearly independent and so the set

$$X = \{\mathbf{0}, \mathbf{H}_i, \mathbf{H}_j + \mathbf{H}_k \mid 1 \leq i \leq 90, 1 \leq j < k \leq 90\}$$

is a set of $1 + 90 + \dbinom{90}{2}$ *distinct* column vectors. But $1 + 90 +$ $\dbinom{90}{2} = 2^{12}$ and so X is precisely the set $V(12, 2)$ of all binary 12-tuples. Hence the number of vectors of odd weight in X is 2^{11} (see e.g. Exercise 2.4 or Exercise 5.5). We now calculate this number in a different way. Suppose m of the columns of H have odd weight, so that $90 - m$ of them have even weight. As in Lemma 2.6, $w(\mathbf{H}_j + \mathbf{H}_k) = w(\mathbf{H}_j) + w(\mathbf{H}_k) - 2w(\mathbf{H}_j \cap \mathbf{H}_k)$, and so $w(\mathbf{H}_j + \mathbf{H}_k)$ is odd if and only if exactly one of $w(\mathbf{H}_j)$ and $w(\mathbf{H}_k)$ is odd. Thus another expression for the number of vectors of odd weight in X is $m + m(90 - m)$. Hence

$$m(91 - m) = 2^{11}$$

and so both m and $91 - m$ are powers of 2. This is clearly impossible for any integer m and so the desired linear code cannot exist.

Remark The non-existence of a non-linear $(90, 2^{78}, 5)$-code will be demonstrated in Theorem 9.7.

The binary Golay [23, 12, 7]-code

We present here the binary Golay code, as did Golay in his 1949 paper, by exhibiting a generator matrix. This is a little unsatisfactory in that it is not clear where the matrix has come from, but it should at least satisfy the reader that the code exists (it will be defined in a more natural way, as a cyclic code, in Chapter 12). Following the treatment of Pless (1982) and MacWilliams and Sloane (1977), we give a different, though equivalent, generator matrix from that given by Golay in order to facilitate the derivation of the code's properties and particularly its minimum distance.

By Theorem 2.7 and Exercise 5.4, the existence of a $[23, 12, 7]$-code C implies the existence of a $[24, 12, 8]$-code $\hat{C}$ and vice versa. It turns out to be advantageous to define the *extended Golay code $\hat{C}$* first.

Theorem 9.3 The code G_{24} having generator matrix $G = [I_{12} \mid A]$

$$
= \begin{bmatrix}
1 & & & & & & & & & & & & 0 & 1 & 1 & 1 & 1 & 1 & 1 & 1 & 1 & 1 & 1 & 1 \\
& 1 & & & & & & & & & & & 1 & 1 & 1 & 0 & 1 & 1 & 1 & 0 & 0 & 0 & 1 & 0 \\
& & 1 & & & & & & & & & & 1 & 1 & 0 & 1 & 1 & 1 & 0 & 0 & 0 & 1 & 0 & 1 \\
& & & 1 & & & & 0 & & & & & 1 & 0 & 1 & 1 & 1 & 0 & 0 & 0 & 1 & 0 & 1 & 1 \\
& & & & 1 & & & & & & & & 1 & 1 & 1 & 1 & 0 & 0 & 0 & 1 & 0 & 1 & 1 & 0 \\
& & & & & 1 & & & & & & & 1 & 1 & 1 & 0 & 0 & 0 & 1 & 0 & 1 & 1 & 0 & 1 \\
& & & & & & 1 & & & & & & 1 & 1 & 0 & 0 & 0 & 1 & 0 & 1 & 1 & 0 & 1 & 1 \\
& & & & & & & 1 & & & & & 1 & 0 & 0 & 0 & 1 & 0 & 1 & 1 & 0 & 1 & 1 & 1 \\
& 0 & & & & & & & 1 & & & & 1 & 0 & 0 & 1 & 0 & 1 & 1 & 0 & 1 & 1 & 1 & 0 \\
& & & & & & & & & 1 & & & 1 & 0 & 1 & 0 & 1 & 1 & 0 & 1 & 1 & 1 & 0 & 0 \\
& & & & & & & & & & 1 & & 1 & 1 & 0 & 1 & 1 & 0 & 1 & 1 & 1 & 0 & 0 & 0 \\
& & & & & & & & & & & 1 & 1 & 0 & 1 & 1 & 0 & 1 & 1 & 1 & 0 & 0 & 0 & 1 \\
\end{bmatrix}
$$

is a $[24, 12, 8]$-code.

Proof We are required to show that $d(G_{24}) = 8$, and by Theorem 5.2 it is enough to show that every non-**0** codeword has weight at least 8. The above generator matrix has been chosen so that this can be done without having to list all 2^{12} codewords. We proceed by a sequence of four lemmas.

Lemma 1 $G_{24}^\perp = G_{24}$, i.e. G_{24} is *self-dual*.

Proof It is readily checked that $\mathbf{u} \cdot \mathbf{v} = 0$, or equivalently that $w(\mathbf{u} \cap \mathbf{v})$ is even, for every pair of (not necessarily distinct) rows $\mathbf{u}$ and $\mathbf{v}$ of G. (The amount of checking involved here can be much reduced by observing that each of rows 3 to 12 of matrix A can be obtained from the second row by means of a cyclic shift of the last 11 coordinates. For, by symmetry arguments, it is then sufficient to calculate $w(\mathbf{u} \cap \mathbf{v})$ only for those pairs of rows $\mathbf{u}$, $\mathbf{v}$ of G in which $\mathbf{u}$ is one of the first two rows.) Hence, each row of G is orthogonal to all the rows of G and so, by Lemma 7.2, $G_{24} \subseteq G_{24}^\perp$. But, by Theorem 7.3, G_{24} and $G_{24}^\perp$ both have dimension 12 and so $G_{24} = G_{24}^\perp$.

Lemma 2 $[A \mid I]$ is also a generator matrix for G_{24}.

Proof By Theorem 7.6, $G_{24}^\perp$ has generator matrix $[A^T \mid I]$, and so the result follows from Lemma 1 and the observation that $A^T = A$.

Lemma 3 Every codeword of G_{24} has weight divisible by 4.

Proof If $\mathbf{u}$ and $\mathbf{v}$ are any two codewords of G_{24}, then $w(\mathbf{u} \cap \mathbf{v}) \equiv \mathbf{u} \cdot \mathbf{v} \equiv 0 \pmod 2$, since G_{24} is self-dual. Observe that all the rows of G have weight divisible by 4. Let $\mathbf{u}$ and $\mathbf{v}$ be two such rows. Then, by Lemma 2.6, $w(\mathbf{u} + \mathbf{v}) = w(\mathbf{u}) + w(\mathbf{v}) - 2w(\mathbf{u} \cap \mathbf{v})$, and since we have just shown that $w(\mathbf{u} \cap \mathbf{v})$ is divisible by 2, it follows that $w(\mathbf{u} + \mathbf{v})$ is divisible by 4. The same argument, with $\mathbf{u}$ a row of G and $\mathbf{v}$ the sum of two rows of G, shows that the sum of any three rows of G has weight divisible by 4, and so on. Thus every linear combination of rows of G has weight divisible by 4.

Lemma 4 G_{24} has no codewords of weight 4.

Proof We write a codeword $\mathbf{x} = x_1 x_2 \cdots x_{24}$ as $(\mathbf{L} \mid \mathbf{R})$ where $\mathbf{L} = x_1 \cdots x_{12}$ is the left half of $\mathbf{x}$ and $\mathbf{R} = x_{13} \cdots x_{24}$ is the right half of $\mathbf{x}$. Suppose $\mathbf{x}$ is a codeword of G_{24} of weight 4. Then one of the following cases occurs.

Case 1 $w(\mathbf{L}) = 0$, $w(\mathbf{R}) = 4$. This is impossible since we see from the generator matrix G that $\mathbf{0}$ is the only codeword with $w(\mathbf{L}) = 0$.

Case 2 $w(\mathbf{L}) = 1$, $w(\mathbf{R}) = 3$. If $w(\mathbf{L}) = 1$, then $\mathbf{x}$ is one of the rows of G, none of which has $w(\mathbf{R}) = 3$.

Case 3 $w(\mathbf{L}) = 2$, $w(\mathbf{R}) = 2$. If $w(\mathbf{L}) = 2$, then $\mathbf{x}$ is the sum of two rows of G, but it is easily seen that no sum of two rows of A has weight 2.

Case 4 $w(\mathbf{L}) = 3$, $w(\mathbf{R}) = 1$. It would be tedious to check that the sum of any three rows of G has $w(\mathbf{R}) > 1$. But by using Lemma 2 we can avoid this. For if $w(\mathbf{R}) = 1$, then $\mathbf{x}$ must be one of the rows of $[A \mid I]$, none of which has weight 4.

Case 5 $w(\mathbf{L}) = 4$, $w(\mathbf{R}) = 0$. Again by looking at the generator matrix $[A \mid I]$ we see that $\mathbf{0}$ is the only codeword having $w(\mathbf{R}) = 0$.

Theorem 9.3 now follows immediately from Lemmas 3 and 4.

The *binary Golay code* G_{23} is obtained from G_{24} simply by omitting the last coordinate position from all codewords. G_{23} is thus a $(23, 2^{12}, 7)$-code whose parameters satisfy the sphere-packing condition

i.e. $2^{12}\left\{1 + 23 + \binom{23}{2} + \binom{23}{3}\right\} = 2^{23}.$

So G_{23} is a perfect code.

Remark The omission of any other fixed coordinate from G_{24} (this process is called *puncturing*) would also give a $(23, 2^{12}, 7)$-code and it happens that any such punctured code is equivalent to G_{23}.

The ternary Golay [11, 6, 5]-code

With just a little trial and error it is not difficult to make use of Theorem 8.4 and to construct the parity-check matrix of an $[11, 6, 5]$-code over $GF(3)$ (see Exercise 9.3).

However, to bring out the similarities of the binary and ternary Golay codes, we exhibit a generator matrix for a ternary $[12, 6, 6]$-code G_{12}, which may be punctured to get the perfect *ternary Golay code G_{11}* with parameters $[11, 6, 5]$.

Theorem 9.4 The ternary code G_{12} having generator matrix

$$
G = [I_6 \mid A] = \begin{bmatrix}
1 & & & & & & 0 & 1 & 1 & 1 & 1 & 1 \\
 & 1 & & & 0 & & 1 & 0 & 1 & 2 & 2 & 1 \\
 & & 1 & & & & 1 & 1 & 0 & 1 & 2 & 2 \\
 & & & 1 & & & 1 & 2 & 1 & 0 & 1 & 2 \\
 & & 0 & & 1 & & 1 & 2 & 2 & 1 & 0 & 1 \\
 & & & & & 1 & 1 & 1 & 2 & 2 & 1 & 0
\end{bmatrix}
$$

is a $[12, 6, 6]$-code.

Proof This is left to Exercise 9.2.

Are there any more perfect codes?

It was conjectured for some time that the Hamming codes Ham (r, q) and the Golay codes G_{23} and G_{11} were the only non-trivial perfect codes. However, in 1962, J. L. Vasil'ev constructed a family of non-linear perfect codes with the same parameters as the binary Hamming codes. Then Schönheim (1968) and Lindström (1969) gave non-linear codes with the same parameters as Hamming codes over $GF(q)$ for any prime power q.

    The conjecture was weakened to: 'any non-trivial perfect code has the parameters of a Hamming or Golay code'. The proof of this, for q a prime power, was finally completed by Tietäväinen (1973) following major contributions by van Lint (see van Lint (1975)). Thus we have the following result, which was also proved independently by Zinov'ev and Leont'ev (1973).

Theorem 9.5 (van Lint and Tietäväinen) A non-trivial perfect q-ary code, where q is a prime power, must have the same parameters as one of the Hamming or Golay codes.

The proof of Theorem 9.5 is rather complicated and the details, which may be found in MacWilliams and Sloane (1977), are omitted here. One important ingredient of the proof is Lloyd's theorem, which we also state without proof, which gives a further necessary condition on the parameters for the existence of a perfect code. The binomial coefficient $\binom{x}{m}$ in the following is defined by

$$\binom{x}{m} = \frac{x(x-1)\cdots(x-m+1)}{m!} \qquad \text{if } m \text{ is a positive integer}$$

$$= 1 \qquad\qquad\qquad\qquad \text{if } m = 0.$$

Theorem 9.6 (Lloyd (1957))   If there exists a perfect $(n, M, 2t + 1)$-code over $GF(q)$, then the polynomial $L_t(x)$ defined by

$$L_t(x) = \sum_{j=0}^{t} (-1)^j (q-1)^{t-j} \binom{x-1}{j}\binom{n-x}{t-j}$$

has t distinct integer roots in the interval $1 \leqslant x \leqslant n$.

Using Lloyd's theorem, it was shown that an unknown perfect code over $GF(q)$ must have $t \leqslant 11$, $q \leqslant 8$ and $n < 485$. However, by the computer search mentioned earlier, the only parameters in this range satisfying the sphere-packing condition are those of trivial, Hamming or Golay codes and also the parameters $(n, M, d) = (90, 2^{78}, 5)$ with $q = 2$. [*Remark*: It has been shown by H. W. Lenstra and A. M. Odlyzko (unpublished) that the computer search can be avoided by tightening the inequalities.]

We have already established the non-existence of a linear $(90, 2^{78}, 5)$-code. The non-existence of a non-linear code with these parameters follows from Lloyd's theorem, for with $t = 2$ and $n = 90$,

$$L_2(x) = 0 \quad \text{if and only if} \quad x^2 - 91x + 2048 = 0$$

and this equation does not have integer solutions in x.

We give below a self-contained proof of this non-existence, avoiding Lloyd's theorem, and relying only on a simple counting argument. We first give a simple definition.

Definition If **u** and **v** are binary vectors of the same length, then we say that **u** *covers* **v** if the 1s in **v** are a subset of the 1s in **u**. In other words,

$$\mathbf{u} \text{ covers } \mathbf{v} \quad \text{if and only if} \quad \mathbf{u} \cap \mathbf{v} = \mathbf{v}.$$

For example 111001 covers 101000.

Theorem 9.7 There does not exist a binary $(90, 2^{78}, 5)$-code.

Proof Suppose, for a contradiction, that C is a $(90, 2^{78}, 5)$-code. By Lemma 2.3, we may assume that $\mathbf{0} \in C$. Then every non-zero codeword in C has weight at least 5. Let Y be the set of vectors in $V(90, 2)$ of weight 3 which begin with two 1s. Clearly $|Y| = 88$. Since C is perfect, each vector $\mathbf{y}$ of Y lies in a unique sphere $S(\mathbf{x}, 2)$ of radius 2 about some codeword $\mathbf{x}$. Such a codeword $\mathbf{x}$ must have weight 5 and must cover $\mathbf{y}$.

Let X be the set of all codewords of C of weight 5 which begin with two 1s. We will count in two ways the number of ordered pairs in the set

$$D = \{(\mathbf{x}, \mathbf{y}) \mid \mathbf{x} \in X, \mathbf{y} \in Y, \mathbf{x} \text{ covers } \mathbf{y}\}.$$

By the previous remarks, each $\mathbf{y}$ in Y is covered by a unique $\mathbf{x}$ in X and so

$$|D| = |Y| = 88.$$

On the other hand, each $\mathbf{x}$ in X (e.g. $1111100\cdots 0$) covers exactly three **y**s in Y ($111000\cdots 0, 110100\cdots 0$ and $110010\cdots 0$), and so

$$|D| = 3|X|.$$

Hence $3|X| = 88$, giving $|X| = 88/3$, which is a contradiction, since $|X|$ must be an integer. Thus a $(90, 2^{78}, 5)$-code cannot exist.

t-designs

The counting argument, which will be generalized in Exercise 9.5(b), of the proof of Theorem 9.7 is reminiscent of that used in proving the relations (2.24) and (2.25) for block designs (see

Exercise 2.13). This is not just a coincidence, for we can associate with any perfect code a certain design called a *t*-design.

Definition A *t-design* consists of a set X of v *points*, and a collection of distinct k-subsets of X, called *blocks*, with the property that any t-subset of X is contained in exactly λ blocks. We call this a *t-(v, k, λ) design*.

Thus 2-designs are the same as balanced block designs, which were defined in Chapter 2.

Definition A *Steiner system* is a *t*-design with $\lambda = 1$. A $t-(v, k, 1)$ design is usually called an $S(t, k, v)$.

For example, the Fano plane of Example 2.19 is an $S(2, 3, 7)$.

The following theorem shows how Steiner systems can be obtained from perfect codes.

Theorem 9.8 (Assmus and Mattson 1967) If there exists a perfect binary t-error-correcting code of length n, then there exists a Steiner system $S(t + 1, 2t + 1, n)$.

Proof This is left to Exercises 9.4(b) and 9.5.

Assmus and Mattson (1969) later gave an important sufficient condition on a code, which is not necessarily perfect, for the existence of associated t-designs. For the details, see MacWilliams and Sloane (1977, Chapter 6) or Assmus and Mattson (1974). Many new 5-designs have been obtained in this way. [*Remark:* it was a long-standing conjecture that t designs having $t \geq 6$ did not exist; however the discovery of a 6-design has recently been announced by Magliveras and Leavitt (1983).]

Remaining problems on perfect codes

Theorem 9.5 leaves the following problems unresolved.

Problem 9.9 Find all perfect codes having the parameters of the Hamming and Golay codes.

It was observed after the definition of the q-ary Hamming codes in Chapter 8 that any *linear* code with the Hamming parameters is equivalent to a Hamming code. But the problem of finding all non-linear codes with these parameters appears to be very difficult and is unsolved. It is believed that there are (at least) several thousand inequivalent perfect binary codes with the parameters $(15, 2^{11}, 3)$. For supporting evidence see Phelps (1983).

However, the two perfect Golay codes are unique, i.e. any code with the parameters of a Golay code must be equivalent to a Golay code. This was proved by Pless (1968) in the restriction to linear codes (see also Exercise 9.3 for the ternary case). For unrestricted codes, the uniqueness of G_{23} was proved by Snover (1973), while that of both G_{23} and G_{11} was demonstrated by Delsarte and Goethals (1975).

Problem 9.10 Find all perfect codes over non-prime-power alphabets.

It is conjectured that there are no non-trivial perfect codes over non-prime-power alphabets. The best result to date is the following theorem of Best (1982), the proof of which is too involved to include here. For an outline, see Best (1983).

Theorem 9.11 For $t \geq 3$ and $t \neq 6$ or 8, the only non-trivial perfect t-error-correcting code over any alphabet is the binary Golay code.

It is likely that the cases $t = 6$ and $t = 8$ (and possibly even $t = 2$) will be settled fairly soon†, but for $t = 1$, the problem appears to be extremely difficult. We have already observed that the parameters

$$(n, M, d) = ((q^r - 1)/(q - 1), q^{n-r}, 3)$$

satisfy the sphere-packing condition for integers q and $r \geq 2$. For q a prime-power, these are the parameters of the Hamming codes, but for q not a prime power, very little is known about the existence or otherwise of codes having these parameters; only in the smallest case, $q = 6$, $r = 2$, is the problem resolved, as we now describe.

The possible existence of a 6-ary $(7, 6^5, 3)$-code was first

---

† Cases $t = 6, 8$ have now been settled by Y. Hong (Ph.D. Dissertation, Ohio State University, 1984).

considered explicitly by Golay (1958) and answered in the negative by Golomb and Posner (1964), who reduced the problem to one from recreational mathematics, posed by Euler in 1782 and solved in 1901, as follows.

Theorem 9.12 There does not exist a 6-ary $(7, 6^5, 3)$-code.

Proof Suppose, for a contradiction, that C is a $(7, 6^5, 3)$-code over the alphabet $F_6 = \{1, 2, 3, 4, 5, 6\}$. Consider the 6^5 vectors of length 5 obtained by deleting the last two coordinates of each codeword of C. These must be precisely the 6^5 distinct vectors of $(F_6)^5$, for if two of these 5-tuples were the same, then the corresponding two codewords in C would be distance at most 2 apart. Hence there are 6^2 codewords of C beginning with any fixed triple. If we now take those 36 codewords of C beginning with 111 and then delete these first three positions, we will have a $(4, 6^2, 3)$-code, which we denote by D. By the same argument as above, the 36 ordered pairs given by deleting any two fixed coordinates from the codewords of D will be precisely the 36 distinct ordered pairs in $(F_6)^2$. Hence, if a codeword $(ijkl)$ of the code D is identified with an officer whose rank is i and whose regiment is j and who stands in the kth row and lth column of a 6×6 square, we have a solution to the following problem:

Euler's '36 officers problem' (1782) There are 36 officers, one from each of 6 ranks from each of 6 regiments. Can these officers be arranged in a 6×6 square so that every row and every column of the square contains one officer of each rank and one officer of each regiment?

It was conjectured by Euler that the answer is 'no', and this was proved to be the case (by exhaustive search) by Tarry (1901). For a fairly short, self-contained proof, see Stinson (1984).

Hence a 6-ary $(7, 6^5, 3)$-code cannot exist and Theorem 9.12 is proved.

Remark The '36 officers problem' is equivalent to a problem concerning mutually orthogonal Latin squares, a topic whose connection with codes is the subject of the next chapter, where it will be seen why the method of proof of Theorem 9.12 cannot be used to rule out the existence of q-ary $(q + 1, q^{q-1}, 3)$-codes for values of q other than 6.

Concluding remarks

(1) The Golay codes have been constructed in a number of different ways, most naturally as cyclic codes (see Chapter 12) or as quadratic residue codes. A less obvious, but neat elementary construction is given in van Lint (1982).

(2) A number of special algorithms have been devised for decoding G_{23} and G_{24}, some of them making ingenious use of the properties of the associated 5-design. Among these are Berlekamp's (1972) algorithm, Goethals' (1971) majority logic algorithm, and Gibson and Blake's (1978) method using 'miracle octad generators'.

(3) The probability of error correction when using G_{23} was found in Exercise 6.6. By Exercise 9.1, there is no advantage in using G_{24} rather than G_{23} for complete decoding.

Exercises 9

9.1   (Generalization of Exercise 8.4)   Suppose C is a perfect binary linear code of length n and that $\hat{C}$ is its extended code. Prove that, for a binary symmetric channel,

$$P_{\text{corr}}(C) = P_{\text{corr}}(\hat{C}).$$

$\Big[$*Hint:* Use the Pascal identity for binomial coefficients,

$$\binom{n + 1}{i} = \binom{n}{i} + \binom{n}{i - 1} \qquad \text{for } n \geq i \geq 1.\Big]$$

9.2   Prove Theorem 9.4; i.e. show that $d(G_{12}) = 6$. [*Hint:* Show that $G_{12}^{\perp} = G_{12}$, so that $[-A^T \mid I]$ is also a generator matrix for G_{12}. Then use the fact that if $w(\mathbf{x}) \leq 5$, then either $w(\mathbf{L}) \leq 2$ or $w(\mathbf{R}) \leq 2$, where $\mathbf{x} = (\mathbf{L} \mid \mathbf{R})$].

9.3   Use Theorem 8.4 to construct G_{11}; i.e. find 11 vectors of $V(5, 3)$ such that any 4 of them are linearly independent. Furthermore show that this can be done in essentially only one way, thus proving the uniqueness of G_{11} as a linear [11, 6, 5]-code. [*Hint:* Show first that, up to equivalence, we may assume that $H = [I_5 \,|\, A]$, where

$$A = \begin{bmatrix} 1 & 1 & 1 & 1 & 1 & 0 \\ 1 & * & * & * & 0 & * \\ 1 & * & * & 0 & * & * \\ 1 & * & 0 & * & * & * \\ 1 & 0 & * & * & * & * \end{bmatrix}$$

and the asterisks represent non-zero entries.]

9.4 (a) Show that if **y** is a vector in $V(23, 2)$ of weight 4, then there exists a unique codeword **x** of weight 7 in G_{23} which covers **y**. Deduce that the number of codewords of weight 7 in G_{23} is 253.

(b) Let M be a matrix whose columns are the codewords of weight 7 in G_{23}. Show that M is the incidence matrix of a design which has 23 points, 253 blocks, 7 points in each block, and such that any 4 points lie together in exactly one block; thus we have constructed a Steiner system $S(4, 7, 23)$.

9.5 Show that if there exists a perfect binary t-error-correcting code of length n, then

(a) there exists a Steiner system $S(t + 1, 2t + 1, n)$;

(b) $\dbinom{n-i}{t+1-i} \Big/ \dbinom{2t+1-i}{t+1-i}$ is an integer for $i = 0, 1, \ldots, t$.

[*Remark:* Putting $n = 90$, $t = 2$ and $i = 2$ in part (b) is the case considered in proving Theorem 9.7.]

9.6 Construct a Steiner system $S(5, 8, 24)$ from the extended binary Golay code G_{24}.

9.7 Show that the number of codewords of weight 3 in the Hamming code Ham $(r, 2)$ is $(2^r - 1)(2^{r-1} - 1)/3$.

9.8 Show that the number of vectors of weight 5 in the ternary Golay code is 132.

9.9 We shall construct the Nordstrom–Robinson $(15, 256, 5)$-code N_{15} in the following steps.

(i) Show that if the order of the coordinates of the binary Golay code G_{24} is changed so that one of the weight 8 codewords is $1111111100 \cdots 0$, then G_{24} has a generator matrix having its first 8 columns as shown below.

$$
G = \begin{bmatrix}
1 & & & & & & 1 & \\
 & 1 & & & 0 & & 1 & \\
 & & 1 & & & & 1 & \\
 & & & 1 & & & 1 & \\
 & & & & 1 & & 1 & \\
 & 0 & & & & 1 & 1 & \\
 & & & & & 1 & 1 & \\
 & & & & & & 0 & \\
 & & & & & & 0 & \\
 & & 0 & & & 0 & & \\
 & & & & & & 0 & \\
 & & & & & & 0 & \\
\end{bmatrix}.
$$

[*Hint:* Since G_{24} is self-dual, (a) the first seven columns of G must be linearly independent and (b) the codeword $1111111100\cdots0$ is orthogonal to every codeword.]

(ii) Show that the total number of codewords of G_{24} whose first eight coordinates are one of 00000000, 10000001, 01000001, 00100001, 00010001, 00001001, 00000101 or 00000011 is 256.

(iii) Take these 256 codewords and delete the first 8 coordinates of each of them. Show that the resulting code is a $(16, 256, 6)$-code. This is the extended Nordstrom–Robinson code N_{16}.

(iv) Puncture N_{16} (e.g. delete the last coordinate) to get the $(15, 256, 5)$-code N_{15}.
      [*Remark:* N_{16} and N_{15} are non-linear codes. They are both optimal, cf. Table 2.4.]

9.10 Construct from N_{15} a $(12, 32, 5)$-code. [This code is called the *Nadler code*, having originally been constructed in another way by Nadler (1962). The Nadler code is both optimal (see Chapter 17, §4 of MacWilliams and Sloane 1977) and unique (Goethals 1977).]

9.11 (i) Show that there does not exist a binary *linear*

(13, 64, 5)-code. [*Hint:* Suppose C is a binary [13, 6, 5]-code with generator matrix

$$\begin{bmatrix} 1\ 1\ 1\ 1\ 1 & 0\ 0\ 0\ 0\ 0\ 0\ 0\ 0 \\ G_1 & G_2 \end{bmatrix}.$$

Show that G_2 generates an [8, 5, 3]-code, whose parameters violate the sphere-packing condition.]
- (ii) Deduce that there is no linear code with the parameters of the Nordstrom–Robinson code.
- (iii) Can the non-existence of a [12, 5, 5]-code (i.e. a linear code with the parameters of the Nadler code) be proved by the method of (i)?

10 Codes and Latin squares

The main aim of this chapter is to show how codes can be constructed from certain sets of Latin squares and vice versa. In particular, we shall completely solve the 'main coding theory problem', over any alphabet, for single-error-correcting codes of length 4.

Latin squares

Definition A *Latin square of order q* is a $q \times q$ array whose entries are from a set F_q of q distinct symbols such that each row and each column of the array contains each symbol exactly once.

Example Let $F_3 = \{1, 2, 3\}$. Then an example of a Latin square of order 3 is

$$
\begin{array}{ccc}
1 & 2 & 3 \\
2 & 3 & 1 \\
3 & 1 & 2.
\end{array}
$$

Latin squares, like balanced block designs (see Chapter 2), can be used in statistical experiments.

Example 10.1 Three headache drugs 1, 2, 3 are to be tested on subjects A, B, C on three successive days M, T, W. One possible schedule is

	M	T	W
A	1	2	3
B	1	2	3
C	1	2	3.

But in addition to testing the effect of different drugs on the same subject, we also want to have some measurement of the effects of the drugs when taken on different days of the three-day period. So we would like each drug to be used exactly once each day, i.e.

we require a Latin square for the schedule, e.g.

$$
\begin{array}{c c c c}
 & M & T & W \\
A & 1 & 2 & 3 \\
B & 2 & 3 & 1 \\
C & 3 & 1 & 2.
\end{array}
$$

Theorem 10.2 There exists a Latin square of order q for any positive integer q.

Proof We can take $1\,2\cdots q$ as the first row and cycle this round once for each subsequent row to get

$$
\begin{array}{l}
1\ 2\ 3\ \cdots \qquad q \\
2\ 3\ 4\ \cdots \quad q\ 1 \\
3\ 4\ 5\ \cdots\ q\ 1\ 2 \\
\vdots\ \vdots\ \vdots \qquad\quad \vdots \\
q\ 1\ 2\ \cdots \qquad q-1.
\end{array}
$$

Alternatively, the addition table of Z_q is a Latin square of order q.

Mutually orthogonal Latin squares

Definition Let A and B be two Latin squares of order q. Let a_{ij} and b_{ij} denote the i, jth entries of A and B respectively. Then A and B are said to be *mutually orthogonal* Latin squares (abbreviated to MOLS) if the q^2 ordered pairs $(a_{ij}, b_{ij}), i, j = 1, 2, \ldots, q$, are all distinct.

In other words, if we superimpose the two squares to form a new $q \times q$ square with ordered pairs as entries, then these q^2 ordered pairs are all distinct.

Example 10.3 The Latin squares

$$
A = \begin{array}{c c c}
1 & 2 & 3 \\
2 & 3 & 1 \\
3 & 1 & 2
\end{array}
\qquad \text{and} \qquad
B = \begin{array}{c c c}
1 & 2 & 3 \\
3 & 1 & 2 \\
2 & 3 & 1
\end{array}
$$

form a pair of MOLS of order 3, for when superimposed they give the array

$$(1,1) \quad (2,2) \quad (3,3)$$
$$(2,3) \quad (3,1) \quad (1,2)$$
$$(3,2) \quad (1,3) \quad (2,1).$$

Application Suppose three headache drugs, labelled 1, 2, 3, and three fever drugs, also labelled 1, 2, 3, are to be tested on three subjects A, B, C on three successive days M, T, W. As in Example 10.1, we shall use a Latin square of order 3 for the headache drug schedule and another one for the fever drug schedule. Since each subject takes a headache drug and a fever drug each day we have the opportunity of observing their combined effect. Can we test each of the 9 combinations of headache drug/fever drug exactly once? Yes, by using the above pair of MOLS.

	M	T	W
A	$(1,1)$	$(2,2)$	$(3,3)$
B	$(2,3)$	$(3,1)$	$(1,2)$
C	$(3,2)$	$(1,3)$	$(2,1)$

Here (i,j) denotes (headache drug i, fever drug j).

Example 10.4 There does not exist a pair of MOLS of order 2, for if $F_2 = \{1, 2\}$, then the only Latin squares of order 2 are $\begin{smallmatrix}1 & 2\\ 2 & 1\end{smallmatrix}$ and $\begin{smallmatrix}2 & 1\\ 1 & 2\end{smallmatrix}$, and these are not mutually orthogonal.

Optimal single-error-correcting codes of length 4

Over an arbitrary alphabet F_q, let us consider the 'main coding theory problem' for codes of length 4 and minimum distance 3; i.e. the problem of finding the value of $A_q(4, 3)$. First we find an upper bound.

Theorem 10.5 $A_q(4, 3) \leqslant q^2$, for all q.

Proof Suppose C is a q-ary $(4, M, 3)$-code and let $\mathbf{x} = x_1x_2x_3x_4$ and $\mathbf{y} = y_1y_2y_3y_4$ be distinct codewords of C. Then $(x_1, x_2) \neq (y_1, y_2)$, for otherwise $\mathbf{x}$ and $\mathbf{y}$ could differ only in the last two places, contradicting $d(C) = 3$. Thus the M ordered pairs obtained by deleting the last two coordinates from C are all distinct vectors of $(F_q)^2$ and so we must have $M \leq q^2$.

Example 10.6 For $q = 3$, the bound of Theorem 10.5 is attained, for the Hamming code Ham $(2, 3)$ is a $(4, 9, 3)$-code:

$$
\begin{array}{cccc}
0 & 0 & 0 & 0 \\
0 & 1 & 1 & 2 \\
0 & 2 & 2 & 1 \\
1 & 0 & 1 & 1 \\
1 & 1 & 2 & 0 \\
1 & 2 & 0 & 2 \\
2 & 0 & 2 & 2 \\
2 & 1 & 0 & 1 \\
2 & 2 & 1 & 0.
\end{array}
$$

Note that the ordered pairs in *any* two fixed coordinate positions are precisely the distinct vectors of $(F_3)^2$. The argument of the proof of Theorem 10.5 shows that this must be so.

Remark For $q \geq 4$, the bound of Theorem 10.5 is a big improvement on the sphere-packing bound, which gives only that $A_q(4, 3) \leq q^4/(4q - 3)$.

Our next task is to determine those values of q for which a q-ary $(4, q^2, 3)$-code exists. Since the q^2 ordered pairs starting off the codewords of such a code are distinct, such a code must have the form

$$\{(i, j, a_{ij}, b_{ij}) \mid (i, j) \in (F_q)^2\}.$$

We now demonstrate the connection between such codes and pairs of mutually orthogonal Latin squares.

Theorem 10.7 There exists a q-ary $(4, q^2, 3)$-code if and only if there exists a pair of MOLS of order q.

Proof We will show that a code

$$C = \{(i, j, a_{ij}, b_{ij}) \mid (i, j) \in (F_q)^2\}$$

is a $(4, q^2, 3)$-code if and only if $A = [a_{ij}]$ and $B = [b_{ij}]$ form a pair of MOLS of order q.

As in the proof of Theorem 10.5, the minimum distance of C is 3 if and only if, for each pair of coordinate positions, the ordered pairs appearing in those positions are distinct. Now the q^2 pairs (i, a_{ij}) are distinct and the q^2 pairs (j, a_{ij}) are distinct if and only if A is a Latin square. The q^2 pairs (i, b_{ij}) are distinct and the q^2 pairs (j, b_{ij}) are distinct if and only if B is a Latin square. Finally the q^2 pairs (a_{ij}, b_{ij}) are distinct if and only if A and B are mutually orthogonal.

Theorem 10.7 shows that $A_q(4, 3) = q^2$ if and only if there exists a pair of MOLS of order q. We shall show (in Theorem 10.12) that such a pair of MOLS is easily constructed for three quarters of all cases, or more precisely, whenever $q \equiv 0$, 1, or 3 (mod 4).

Theorem 10.8 If q is a prime power and $q \neq 2$, then there exists a pair of MOLS of order q.

Proof Let F_q be the field $GF(q) = \{\lambda_0, \lambda_1, \ldots, \lambda_{q-1}\}$, where $\lambda_0 = 0$ (if q is prime, we may take $\lambda_i = i$ for each i). Let μ and ν be two distinct non-zero elements of $GF(q)$. Let $A = [a_{ij}]$ and $B = [b_{ij}]$ be $q \times q$ arrays defined by

$$a_{ij} = \lambda_i + \mu\lambda_j \quad \text{and} \quad b_{ij} = \lambda_i + \nu\lambda_j.$$

(The rows and columns of A and B are indexed by $0, 1, \ldots, q - 1$.) We first verify that A and B are Latin squares. If two elements in the same row of A are identical, then we have

$$\lambda_i + \mu\lambda_j = \lambda_i + \mu\lambda_{j'}, \quad \text{i.e.} \quad \mu\lambda_j = \mu\lambda_{j'},$$

implying that $j = j'$, since $\mu \neq 0$. Similarly, if two elements in the same column of A are identical, then we have

$$\lambda_i + \mu\lambda_j = \lambda_{i'} + \mu\lambda_j, \quad \text{i.e.} \quad \lambda_i = \lambda_{i'},$$

implying that $i = i'$. Thus A, and similarly B, are Latin squares. To show that A and B are orthogonal, suppose on the contrary that $(a_{ij}, b_{ij}) = (a_{i'j'}, b_{i'j'})$, i.e. assume that the same ordered pair appears twice in the superposition of the squares. Then

$$\lambda_i + \mu\lambda_j = \lambda_{i'} + \mu\lambda_{j'}$$

and
$$\lambda_i + \nu\lambda_j = \lambda_{i'} + \nu\lambda_{j'},$$

which on subtraction implies that

$$(\mu - v)\lambda_j = (\mu - v)\lambda_{j'}.$$

Since $\mu \neq v$, we have $j = j'$ and, consequently, $i = i'$.

Remark Notice how the important field property of being able to cancel non-zero factors was used in the above proof. A similar construction using Z_n, where n is not a prime, would fail to give a pair of MOLS.

Example 10.9 With $GF(3) = \{0, 1, 2\}$, the construction of Theorem 10.8 gives, taking $\mu = 1$, $v = 2$,

$$
A = \begin{matrix} 0 & 1 & 2 \\ 1 & 2 & 0 \\ 2 & 0 & 1 \end{matrix}
\quad \text{and} \quad
B = \begin{matrix} 0 & 2 & 1 \\ 1 & 0 & 2 \\ 2 & 1 & 0. \end{matrix}
$$

The corresponding $(4, 9, 3)$-code, given by Theorem 10.7, is precisely the Hamming code as displayed in Example 10.6.

We next describe a construction which yields pairs of MOLS of order q for many more values of q.

Theorem 10.10 If there exists a pair of MOLS of order m and there exists a pair of MOLS of order n, then there exists a pair of MOLS of order mn.

Proof Suppose A_1, A_2 is a pair of MOLS of order m and B_1, B_2 is a pair of MOLS of order n.

Denote the (i, j)th entry of A_k by $a_{ij}^{(k)}$ $(k = 1, 2)$
     and the (i, j)th entry of B_k by $b_{ij}^{(k)}$ $(k = 1, 2)$.
Let C_1 and C_2 be the $mn \times mn$ squares defined by

$$
\begin{matrix}
C_k = (a_{11}^{(k)}, B_k)(a_{12}^{(k)}, B_k) \cdots (a_{1m}^{(k)}, B_k) \\
(a_{21}^{(k)}, B_k) \\
\vdots \qquad\qquad\qquad \vdots \\
(a_{m1}^{(k)}, B_k) \quad \cdots \quad (a_{mm}^{(k)}, B_k)
\end{matrix}
$$

where $(a_{ij}^{(k)}, B_k)$ denotes an $n \times n$ array whose r, sth entry is $(a_{ij}^{(k)}, b_{rs}^{(k)})$ for $r, s = 1, 2, \ldots, n$.

In other words, C_k is obtained from A_k by replacing each entry

a of A_k by the $n \times n$ array (a, B_k), where

$$(a, B_k) = (a, b_{11}^{(k)})(a, b_{12}^{(k)}) \cdots (a, b_{1n}^{(k)})$$
$$(a, b_{21}^{(k)})$$
$$\vdots \qquad \qquad \vdots$$
$$(a, b_{n,1}^{(k)}) \cdots (a, b_{n,n}^{(k)}).$$

It is a straightforward exercise to verify that C_1 and C_2 are Latin squares and that they are mutually orthogonal.

Example 10.11

$$A_1 = \begin{matrix} 0 & 1 & 2 \\ 1 & 2 & 0 \\ 2 & 0 & 1 \end{matrix} \quad \text{and} \quad A_2 = \begin{matrix} 0 & 1 & 2 \\ 2 & 0 & 1 \\ 1 & 2 & 0 \end{matrix} \quad \begin{matrix} \text{is a pair of} \\ \text{MOLS of order 3.} \end{matrix}$$

$$B_1 = \begin{matrix} 0 & 1 & 2 & 3 \\ 1 & 0 & 3 & 2 \\ 2 & 3 & 0 & 1 \\ 3 & 2 & 1 & 0 \end{matrix} \quad \text{and} \quad B_2 = \begin{matrix} 0 & 1 & 2 & 3 \\ 2 & 3 & 0 & 1 \\ 3 & 2 & 1 & 0 \\ 1 & 0 & 3 & 2 \end{matrix} \quad \begin{matrix} \text{is a pair of} \\ \text{MOLS of order 4.} \end{matrix}$$

The construction of Theorem 10.10 gives the following pair of MOLS of order 12 in which the entries are ordered pairs from the Cartesian product $F_3 \times F_4 = \{00, 01, 02, 03, 10, 11, 12, 13, 20, 21, 22, 23\}$. (We could relabel these elements as the integers $1, 2, \ldots, 12$ if we wished).

$$C_1 = $$

00 01 02 03	10 11 12 13	20 21 22 23
01 00 03 02		
02 03 00 01		
03 02 01 00		
10 11 12 13	20	00
11		
12		
13		
20	00	10
21		
22		
23		

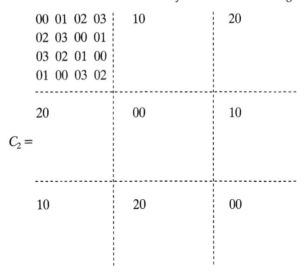

It should be clear to the reader how to complete the remaining entries in the above squares.

The construction of Theorem 10.10 can be repeated any number of times. For example, we can get a pair of MOLS of order 60 by taking the pair of MOLS of order 12 constructed in Example 10.11 together with a pair of MOLS of order 5 as given by Theorem 10.8. The following result tells us precisely for which values of q a pair of MOLS of order q can be constructed by this method.

Theorem 10.12 If $q \equiv 0$, 1 or 3 (mod 4), then there exists a pair of MOLS of order q.

Proof Suppose $q \equiv 0$, 1 or 3 (mod 4). Then q is either odd or is divisible by 4. Hence, if $q = p_1^{h_1} p_2^{h_2} \cdots p_t^{h_t}$ is the prime factorization of q, where $p_1, p_2, \ldots, p_t$ are distinct primes and $h_1, h_2, \ldots, h_t$ are positive integers, then $p_i^{h_i} \geq 3$ for each i. Thus, by Theorem 10.8, there exists a pair of MOLS of order $p_i^{h_i}$ for each i. Repeated application of Theorem 10.10 then gives a pair of MOLS of order $p_1^{h_1} p_2^{h_2} \cdots p_t^{h_t} = q$.

Theorem 10.12 leaves cases $q \equiv 2 \pmod 4$, i.e. $q = 2, 6, 10, 14, \ldots$, unresolved. It was shown in Example 10.4 that there does not exist a pair of MOLS of order 2. A pair of MOLS of order 6 is equivalent to a solution of Euler's '36 officers problem'. As we saw in Chapter 9, Euler's conjecture that no such pair exists was proved by Tarry. Euler conjectured further that there does not exist a pair of MOLS of order q for any $q \equiv 2 \pmod 4$. For such $q \geq 10$, he could not have been further from the truth, though it was not until 1960 that his conjecture was finally disposed of in the following result.

Theorem 10.13 (Bose, Shrikhande and Parker (1960)). There exists a pair of MOLS of order q for *all* q except $q = 2$ and $q = 6$.

The proof of Theorem 10.13 for cases $q \equiv 2 \pmod 4$ is rather complicated and is omitted here.

Corollary 10.14 $A_q(4, 3) = q^2$ for all $q \neq 2, 6$.

Proof This is immediate from Theorems 10.5, 10.7, and 10.13.

Finally we find the values of $A_q(4, 3)$ for $q = 2$ and $q = 6$. It is a very easy exercise to show that $A_2(4, 3) = 2$ (see Exercise 2.1), while the following gives the value of $A_6(4, 3)$.

Theorem 10.15 $A_6(4, 3) = 34$.

Proof The arrays

$$
A = \begin{array}{cccccc}
1 & 2 & 3 & 4 & 5 & 6 \\
2 & 1 & 4 & 3 & 6 & 5 \\
3 & 4 & 6 & 5 & 1 & 2 \\
4 & 3 & 5 & 6 & 2 & 1 \\
5 & 6 & 2 & 1 & 4 & 3 \\
6 & 5 & 1 & 2 & 3 & 4
\end{array}
\quad \text{and} \quad
B = \begin{array}{cccccc}
1 & 2 & 3 & 4 & 5 & 6 \\
3 & 4 & 5 & 6 & 1 & 2 \\
2 & 1 & 4 & 3 & 6 & 5 \\
6 & 5 & 1 & 2 & 4 & 3 \\
4 & 3 & 6 & 5 & 2 & 1 \\
5 & 6 & 2 & 1 & 3 & 4
\end{array}
$$

form a pair of Latin squares which are as close to being orthogonal as is possible. They fail only in that $(a_{65}, b_{65}) = (a_{13}, b_{13})$ and $(a_{66}, b_{66}) = (a_{14}, b_{14})$. Thus the code

$$\{(i, j, a_{ij}, b_{ij}) \mid (i, j) \in (F_6)^2, (i, j) \neq (6, 5) \text{ or } (6, 6)\}$$

is a $(4, 34, 3)$-code.

Now if there existed a $(4, 35, 3)$-code C over F_6, then C would have the form

$$\{(i, j, a_{ij}, b_{ij}) \mid (i, j) \in (F_6)^2, (i, j) \neq (i_0, j_0)\}$$

for some (i_0, j_0). After a little thought, the reader should be able to show that the partial 6×6 arrays $A = [a_{ij}]$ and $B = [b_{ij}]$, each having the (i_0, j_0)th entry missing, can be completed to Latin squares which must be mutually orthogonal. This contradicts Tarry's non-existence result.

Summarizing our results concerning $A_q(4, 3)$, we have

Theorem 10.16 $A_q(4, 3) = q^2$, for all $q \neq 2, 6$,

$$A_2(4, 3) = 2,$$
$$A_6(4, 3) = 34.$$

Remark   We now see why the non-existence of a perfect q-ary $(q + 1, q^{q-1}, 3)$-code cannot be proved by using the method of proof of Theorem 9.12 except when $q = 6$.

In the remainder of this chapter, we generalize some of the earlier results. First we give a generalization of the bound of Theorem 10.5, due to Singleton (1964).

Theorem 10.17   (The Singleton bound)

$$A_q(n, d) \leq q^{n-d+1}.$$

Proof   Suppose C is a q-ary (n, M, d)-code. As in the proof of Theorem 10.5, if we delete the last $d - 1$ coordinates from each codeword (i.e. puncture C $d - 1$ times), then the M vectors of length $n - d + 1$ so obtained must be distinct and so $M \leq q^{n-d+1}$.

Sets of t mutually orthogonal Latin squares

Definition   A set $\{A_1, A_2, \ldots, A_t\}$ of Latin squares of order q is called a set of mutually orthogonal Latin squares (MOLS) if each pair $\{A_i, A_j\}$ is a pair of MOLS, for $1 \leq i < j \leq t$.

Theorem 10.18   There are at most $q - 1$ Latin squares in any set of MOLS of order q.

Proof Suppose $A_1, A_2, \ldots, A_t$ is a set of t MOLS of order q. The orthogonality of two Latin squares is not violated if the elements in any square are relabelled. So we can relabel the elements of each square so that the first row of each A_i is $1\,2\cdots q$. Now consider the t entries appearing in the $(2, 1)$th position of the t Latin squares. None of these entries can be a 1, since 1 already appears in the first column of each A_i. Also, no two of these entries can be the same, because for any two of the A_i, the pairs $(1, 1), (2, 2), \ldots, (q, q)$ already appear in the first row of the corresponding superimposed matrix. Hence we must have $t \le q - 1$.

Definition If a set of $q - 1$ MOLS of order q exists, it is called a *complete* set of MOLS of order q.

Theorem 10.19 If q is a prime power, then there exists a complete set of $q - 1$ MOLS of order q.

Proof Consider the field $GF(q) = \{\lambda_0, \lambda_1, \ldots, \lambda_{q-1}\}$ where $\lambda_0 = 0$. Let $A_1, A_2, \ldots, A_{q-1}$ be $q \times q$ arrays, with rows and columns indexed by $0, 1, \ldots, q - 1$, in which the (i, j)th entry of A_k is the element of $GF(q)$ defined by

$$a_{ij}^{(k)} = \lambda_i + \lambda_k \lambda_j.$$

It follows exactly as in the proof of Theorem 10.8, that $A_1, A_2, \ldots, A_{q-1}$ form a set of MOLS of order q.

Remark It is not known whether there exist any complete sets of MOLS of order q when q is not a prime power. Surprisingly, a complete set of MOLS of order $q \ge 3$ is equivalent to a projective plane of order q (see e.g. Ryser (1963), p. 92 for a proof of this). Thus one approach towards finding a projective plane of order 10 (the lowest-order unsolved case, as mentioned in Chapter 2) is to try to find a set of 9 MOLS of order 10. However, no-one has yet succeeded in finding even a set of 3 MOLS of order 10.

Theorem 10.20 A q-ary $(n, q^2, n - 1)$-code is equivalent to a set of $n - 2$ MOLS of order q.

Proof As in Theorem 10.7, an $(n, q^2, n-1)$-code C over F_q has the form

$$\{(i,j, a_{ij}^{(1)}, a_{ij}^{(2)}, \ldots, a_{ij}^{(n-2)}) \mid (i,j) \in (F_q)^2\}.$$

It is left as an exercise for the reader to show that $d(C) = n-1$ if and only if $A_1, A_2, \ldots, A_{n-2}$, where $A_k = [a_{ij}^{(k)}]$, form a set of MOLS of order q.

Corollary 10.21 $A_q(3, 2) = q^2$ for all q.

Proof A $(3, q^2, 2)$-code is equivalent to a single Latin square of order q, which exists by Theorem 10.2. The Singleton bound shows that such a code is optimal.

Corollary 10.22 If q is a prime power and $n \le q+1$, then

$$A_q(n, n-1) = q^2.$$

Proof This is immediate from Theorems 10.17, 19 and 20.

For other connections between Latin squares and error-correcting codes, see Dénes and Keedwell (1974).

Exercises 10

10.1 Construct a pair of orthogonal Latin squares of order 7.
10.2 Use a pair of MOLS of order 3 and the construction of Theorem 10.10 to construct a pair of MOLS of order 9.
10.3 Using the field $GF(4)$ as defined in Example 3.6(3), construct a set of three MOLS of order 4.
10.4 Show that the dual of the Hamming code Ham $(2, q)$ is a $(q+1, q^2, q)$-code. List the codewords of $(\text{Ham }(2,5))^\perp$ and hence construct a set of four MOLS of order 5.
10.5 Define $f(q)$ to be the largest number of Latin squares in a set of MOLS of order q. On the basis of results stated in this chapter, write down all the information you can about the values of $f(n)$ for $3 \le n \le 20$; i.e. give values of $f(n)$ where known, otherwise give the best upper and lower bounds you can.
10.6 Show that $A_{20}(5, 4) = 400$.

11 A double-error-correcting decimal code and an introduction to BCH codes

In Chapter 3 we met the ISBN code, which is a single-error-*detecting* decimal code of length 10. Then in Example 7.12 we constructed a single-error-*correcting* decimal code of length 10. Our first task in this chapter will be to construct a double-error-correcting decimal code of length 10 and to determine an efficient algorithm for decoding it. As before, the code will really be a linear code defined over $GF(11)$.

We shall then generalize this construction to a family of t-error-correcting codes defined over finite fields $GF(q)$, where $2t + 1 < q$. These codes are particular examples of BCH codes (BCH codes were discovered independently by Hocquenghem (1959) and by Bose and Ray-Chaudhuri (1960)) or Reed–Solomon codes.

We shall see that the decoding of these codes depends on solving a certain system of simultaneous non-linear equations, for which coding theorists have devised some clever methods of solution. Surprisingly, such a system of equations was first solved by Ramanujan (1912) in a seemingly little-known paper in the *Journal of the Indian Mathematical Society*. We shall present here a decoding algorithm based on Ramanujan's method, which is easy to understand and makes use of the method of partial fractions which the reader will very likely have met.

Historical Remark In 1970, N. Levinson wrote an expository article entitled 'Coding Theory—a counterexample to G. H. Hardy's conception of applied mathematics', in which he showed how theorems from number theory play a central role in coding theory, contrary to Hardy's (1940) view that number theory could not have any useful application. It is of particular interest, therefore, to see a result of Hardy's great protegé, Ramanujan, also finding an application in coding theory. Incidentally, perhaps contrary to popular belief, Ramanujan was not completely unknown before his discovery by Hardy. He had already published three papers, all in the above-mentioned journal, before he first wrote to Hardy in January, 1913. It is the third of these papers, published in 1912, which is of interest to us here. It was just two pages long and gave neither references nor any motivation for solving the given system of equations.

Some preliminary results from linear algebra

We shall construct a code of specified minimum distance d by constructing a parity-check matrix H having the property that any $d-1$ columns of H are linearly independent (see Theorem 8.4). The following well-known result concerning the determinant of a Vandermonde matrix enables us to make this construction in a natural way. The determinant of a matrix A will be denoted by $\det A$.

Theorem 11.1 Suppose $a_1, a_2, \ldots, a_r$ are distinct non-zero elements of a field. Then the so-called *Vandermonde matrix*

$$A = \begin{bmatrix} 1 & 1 & \cdots & 1 \\ a_1 & a_2 & \cdots & a_r \\ a_1^2 & a_2^2 & \cdots & a_r^2 \\ \vdots & \vdots & & \vdots \\ a_1^{r-1} & a_2^{r-1} & \cdots & a_r^{r-1} \end{bmatrix}$$

has a non-zero determinant.

Proof By subtracting $a_1 \times$ row i from row $(i+1)$ for $i=1$ to $r-1$, we have

$$\det A = \det \begin{bmatrix} 1 & 1 & \cdots & 1 \\ 0 & a_2 - a_1 & \cdots & a_r - a_1 \\ 0 & a_2(a_2 - a_1) & \cdots & a_r(a_r - a_1) \\ 0 & a_2^2(a_2 - a_1) & \cdots & a_r^2(a_r - a_1) \\ \vdots & \vdots & & \vdots \\ 0 & a_2^{r-2}(a_2 - a_1) & \cdots & a_r^{r-2}(a_r - a_1) \end{bmatrix}$$

$$= (a_2 - a_1)(a_3 - a_1) \cdots (a_r - a_1) \det \begin{bmatrix} 1 & 1 & \cdots & 1 \\ a_2 & a_3 & \cdots & a_r \\ a_2^2 & a_3^2 & \cdots & a_r^2 \\ \vdots & \vdots & & \vdots \\ a_2^{r-2} & a_3^{r-2} & \cdots & a_r^{r-2} \end{bmatrix}.$$

The matrix in this last expression is again of Vandermonde type and if we similarly subtract $a_2 \times$ row i from row $(i + 1)$ for $i = 1$ to $r - 2$, and then take out factors, we get

$$\det A = (a_2 - a_1)(a_3 - a_1) \cdots (a_r - a_1)(a_3 - a_2) \cdots (a_r - a_2)$$

$$\times \det \begin{bmatrix} 1 & 1 & \cdots & 1 \\ a_3 & a_4 & \cdots & a_r \\ \vdots & & & \\ a_3^{r-3} & & \cdots & a_r^{r-3} \end{bmatrix}.$$

Repeating the process until the determinant becomes unity,

$$\det A = (a_2 - a_1)(a_3 - a_1) \cdots (a_r - a_1)$$
$$(a_3 - a_2) \cdots (a_r - a_2)$$
$$\ddots \qquad \vdots$$
$$(a_r - a_{r-1}) \times 1$$
$$= \prod_{i>j} (a_i - a_j).$$

Hence $\det A$ is non-zero, since the a_i are distinct non-zero elements of a field. [*Remark:* the reader who is familiar with the method of proof by induction should be able to shorten the length of the above proof.]

The following is another standard result from linear algebra; its converse is also true, but will not be needed.

Theorem 11.2 If A is an $r \times r$ matrix having a non-zero determinant, then the r columns of A are linearly independent.

Proof Suppose A is an $r \times r$ matrix such that $\det A \neq 0$. Suppose, for a contradiction, that the columns $c_1, c_2, \ldots, c_r$ of A are linearly dependent. Then some column of A can be expressed as a linear combination of the other columns, say

$$c_j = \sum_{\substack{i=1 \\ i \neq j}}^{r} a_i c_i.$$

Then replacing column c_j by $c_j - \sum_{\substack{i=1 \\ i \neq j}}^{r} a_i c_i$ gives a matrix B whose determinant is equal to that of A and which also has an all-zero column. Thus $\det A = \det B = 0$, giving the desired contradiction.

A double-error-correcting modulus 11 code

We are now ready to construct our double-error-correcting decimal code. The code will consist of those codewords of the single-error-correcting code of Example 7.12 which satisfy two further parity-check equations. A similar code was considered by Brown (1974).

Example 11.3 Let C be the linear $[10, 6]$-code over $GF(11)$ defined to have parity-check matrix

$$H = \begin{bmatrix} 1 & 1 & 1 & \cdots & 1 \\ 1 & 2 & 3 & \cdots & 10 \\ 1 & 2^2 & 3^2 & \cdots & 10^2 \\ 1 & 2^3 & 3^3 & \cdots & 10^3 \end{bmatrix}.$$

As usual, if we desire a decimal code rather than one over $GF(11)$, we simply omit those codewords containing the symbol 10 so that our decimal code is

$$D = \left\{ x_1 x_2 \cdots x_{10} \in (F_{10})^{10} \; \middle| \; \sum_{i=1}^{10} x_i \right.$$
$$\left. \equiv \sum_{i=1}^{10} i x_i \equiv \sum_{i=1}^{10} i^2 x_i \equiv \sum_{i=1}^{10} i^3 x_i \equiv 0 \,(\text{mod } 11) \right\}$$

where $F_{10} = \{0, 1, 2, \ldots, 9\}$.

Note that any four columns of H form a Vandermonde matrix and so, by Theorems 11.1 and 11.2, any four columns of H are linearly independent. Thus, by Theorem 8.4, the code C (and hence also D) has minimum distance 5 and so is a double-error-correcting code.

Remark The 11-ary code C contains 11^6 codewords and so is optimal by the Singleton bound (Theorem 10.17). The decimal code D does not achieve the Singleton bound of 10^6 but nevertheless contains over 680 000 codewords.

We next construct a syndrome decoding scheme which will correct all double (and single) errors in codewords of C.

Suppose $\mathbf{x} = x_1 x_2 \cdots x_{10}$ is the transmitted codeword and $\mathbf{y} = y_1 y_2 \cdots y_{10}$ is the received vector. We calculate the syndrome of **y**

$$(S_1, S_2, S_3, S_4) = \mathbf{y}H^T = \left(\sum_{i=1}^{10} y_i, \sum_{i=1}^{10} iy_i, \sum_{i=1}^{10} i^2 y_i, \sum_{i=1}^{10} i^3 y_i \right).$$

Suppose two errors of magnitudes a and b have occurred in positions i and j respectively. Then

$$a + b = S_1 \tag{1}$$
$$ai + bj = S_2 \tag{2}$$
$$ai^2 + bj^2 = S_3 \tag{3}$$
$$ai^3 + bj^3 = S_4. \tag{4}$$

We are required to solve these four equations for the four unknowns a, b, i, j and at first sight this looks rather difficult as the equations are non-linear. However, we can eliminate a, b and j as follows.

$$i \times (1) - (2) \text{ gives } b(i - j) = iS_1 - S_2 \tag{5}$$
$$i \times (2) - (3) \text{ gives } bj(i - j) = iS_2 - S_3 \tag{6}$$
$$i \times (3) - (4) \text{ gives } bj^2(i - j) = iS_3 - S_4. \tag{7}$$

Comparing $(6)^2$ with $(5) \times (7)$ now gives

$$(iS_2 - S_3)^2 = (iS_1 - S_2)(iS_3 - S_4),$$

which implies that

$$(S_2^2 - S_1 S_3)i^2 + (S_1 S_4 - S_2 S_3)i + S_3^2 - S_2 S_4 = 0. \tag{8}$$

It is clear that if a, b and i were eliminated from (1) to (4) in similar fashion, then we would get the same equation (8) with i replaced by j. Thus the error locations i and j are just the roots of the quadratic equation (8). Once i and j are found, the values of a and b are easily obtained from (1) and (2).

Let $P = S_2^2 - S_1 S_3$, $Q = S_1 S_4 - S_2 S_3$ and $R = S_3^2 - S_2 S_4$. Note that if just one error has occurred, say in position i of magnitude a, then we have

$$S_1 = a, \quad S_2 = ai, \quad S_3 = ai^2 \quad \text{and} \quad S_4 = ai^3$$

and so $P = Q = R = 0$.

Thus our decoding algorithm is as follows.

From the received vector **y**, calculate the syndrome $S(\mathbf{y}) = (S_1, S_2, S_3, S_4)$ and, if this is non-zero, calculate P, Q and R.

(i)   If $S(\mathbf{y}) = \mathbf{0}$, then $\mathbf{y}$ is a codeword and we assume no errors.

(ii) If $S(\mathbf{y}) \neq \mathbf{0}$ and $P = Q = R = 0$, then we assume a single error of magnitude S_1 in position S_2/S_1.

(iii) If $P \neq 0$ and $R \neq 0$ and if $Q^2 - 4PR$ is a non-zero square in $GF(11)$, then we assume there are two errors located in positions i and j of magnitudes a and b respectively, where

$$i,j = \frac{-Q \pm \sqrt{(Q^2 - 4PR)}}{2P} \qquad (9)$$

$$b = (iS_1 - S_2)/(i - j) \qquad (10)$$

and

$$a = S_1 - b. \qquad (11)$$

(iv) If none of (i), (ii) or (iii) applies, then we conclude that at least three errors have occurred.

Notes   (1)   It does not matter which way round we take i and j in (9); we need not insist, for example, that $i < j$.

(2)   As usual, all arithmetic is carried out modulo 11, division being carried out with the aid of the table of inverses as in Example 7.12. We need further here a table of square roots modulo 11. By first calculating the squares of the scalars as shown below

x	1 2 3 4 5 6 7 8 9 10
x^2	1 4 9 5 3 3 5 9 4 1

we may take the table of square roots to be

x	1 3 4 5 9
$\sqrt{x}$	1 5 2 4 3

We could equally well use the negative of any of these square roots; the presence of the '$\pm$' in (9) shows that it does not matter which of the two roots is taken. Note that if, in (9), $Q^2 - 4PR$ is not a square (i.e. it is one of 2, 6, 7, 8, 10), then at least three errors must have occurred.

A class of BCH codes

Let us now consider how the code of Example 11.3 might be generalized. Generalizing the construction of the code to a t-error-correcting code of length n over $GF(q)$ is very easy

provided
$$2t + 1 \leqslant n \leqslant q - 1.$$

Generalizing the decoding algorithm is less straightforward but can nevertheless be done in an ingenious way.

The codes defined below belong to the much larger class of BCH codes. By restricting our attention to these easily defined codes we can demonstrate in an elementary way the essential ingredients of the important error-correction procedure for the more general BCH codes.

We will assume for simplicity that q is a prime number, so that $GF(q) = \{0, 1, \ldots, q - 1\}$, but there is no difficulty whatsoever in adapting the results to the general prime-power case.

Let C be the code over $GF(q)$ defined to have the parity-check matrix

$$H = \begin{bmatrix} 1 & 1 & 1 & \cdots & 1 \\ 1 & 2 & 3 & \cdots & n \\ 1 & 2^2 & 3^2 & \cdots & n^2 \\ \vdots & & & & \\ 1 & 2^{d-2} & 3^{d-2} & \cdots & n^{d-2} \end{bmatrix},$$

where $d \leqslant n \leqslant q - 1$. That is,

$$C = \left\{ x_1 x_2 \cdots x_n \in V(n, q) \,\middle|\, \sum_{i=1}^{n} i^j x_i = 0 \text{ for } j = 0, 1, \ldots, d - 2 \right\}.$$

Any $d - 1$ columns of H form a Vandermonde matrix and so are linearly independent by Theorems 11.1 and 11.2. Hence, by Theorem 8.4, C has minimum distance d and so is a q-ary (n, q^{n-d+1}, d)-code. Since C meets the Singleton bound (Theorem 10.17), we have proved

Theorem 11.4 If q is a prime-power and if $d \leqslant n \leqslant q - 1$, then

$$A_q(n, d) = q^{n-d+1}.$$

From now on we will assume that d is odd, so that $d = 2t + 1$ and H has $2t$ rows. Let us try to generalize the decoding algorithm of Example 11.3.

Suppose the codeword $\mathbf{x} = x_1 x_2 \cdots x_n$ is transmitted and that the vector $\mathbf{y} = y_1 y_2 \cdots y_n$ is received in which we assume that at most t errors have occurred. Suppose the errors have occurred in positions $X_1, X_2, \ldots, X_t$ with respective magnitudes $m_1, m_2, \ldots, m_t$ (if $e < t$ errors have occurred, we just assume that $m_{e+1} = m_{e+2} = \cdots = m_t = 0$). From the received vector **y** we calculate the syndrome

$$(S_1, S_2, \ldots, S_{2t}) = \mathbf{y}H^T,$$

i.e. we calculate

$$S_j = \sum_{i=1}^{n} y_i i^{j-1} = \sum_{i=1}^{t} m_i X_i^{j-1}$$

for $j = 1, 2, \ldots, 2t$.

Thus, to find the errors, we must solve for X_i and m_i the following system of equations

$$\left. \begin{array}{llll} m_1 & + m_2 & + \cdots + m_t & = S_1 \\ m_1 X_1 & + m_2 X_2 & + \cdots + m_t X_t & = S_2 \\ m_1 X_1^2 & + m_2 X_2^2 & + \cdots + m_t X_t^2 & = S_3 \\ & & \vdots & \\ m_1 X_1^{2t-1} & + m_2 X_2^{2t-1} & + \cdots + m_t X_t^{2t-1} & = S_{2t}. \end{array} \right\} \quad (11.5)$$

This is precisely the system of equations solved by Ramanujan in 1912 and we follow exactly his method of solution below (for $t \geq 3$, the equations are too complicated to eliminate $2t - 1$ of the unknowns as we did for the case $t = 2$).

Consider the expression

$$\phi(\theta) = \frac{m_1}{1 - X_1 \theta} + \frac{m_2}{1 - X_2 \theta} + \cdots + \frac{m_t}{1 - X_t \theta}. \quad (1)$$

Now

$$\frac{m_j}{1 - X_j \theta} = m_j(1 + X_j \theta + X_j^2 \theta^2 + \cdots)$$

and so

$$\phi(\theta) = (m_1 + m_2 + \cdots + m_t) + (m_1 X_1 + m_2 X_2 + \cdots + m_t X_t)\theta$$
$$+ (m_1 X_1^2 + m_2 X_2^2 + \cdots + m_t X_t^2)\theta^2 + \cdots.$$

By virtue of equations (11.5), we get

$$\phi(\theta) = S_1 + S_2 \theta + S_3 \theta^2 + \cdots + S_{2t} \theta^{2t-1} + \cdots. \quad (2)$$

Reducing the fractions in (1) to a common denominator, we have

$$\phi(\theta) = \frac{A_1 + A_2 \theta + A_3 \theta^2 + \cdots + A_t \theta^{t-1}}{1 + B_1 \theta + B_2 \theta^2 + \cdots + B_t \theta^t}. \quad (3)$$

Hence

$$(S_1 + S_2\theta + S_3\theta^2 + \cdots)(1 + B_1\theta + B_2\theta^2 + \cdots + B_t\theta^t)$$
$$= A_1 + A_2\theta + A_3\theta^2 + \cdots + A_t\theta^{t-1}.$$

Equating like powers of θ we have:

$$\left.\begin{aligned}
A_1 &= S_1 \\
A_2 &= S_2 + S_1 B_1 \\
A_3 &= S_3 + S_2 B_1 + S_1 B_2 \\
&\;\;\vdots \\
A_t &= S_t + S_{t-1}B_1 + S_{t-2}B_2 + \cdots + S_1 B_{t-1}
\end{aligned}\right\} \tag{11.6}$$

$$\left.\begin{aligned}
0 &= S_{t+1} + S_t B_1 + S_{t-1}B_2 + \cdots + S_1 B_t \\
0 &= S_{t+2} + S_{t+1}B_1 + S_t B_2 + \cdots + S_2 B_t \\
&\;\;\vdots \\
0 &= S_{2t} + S_{2t-1}B_1 + S_{2t-2}B_2 + \cdots + S_t B_t.
\end{aligned}\right\} \tag{11.7}$$

Since $S_1, S_2, \ldots, S_{2t}$ are known, the t equations (11.7) enable us to find $B_1, B_2, \ldots, B_t$, and then $A_1, A_2, \ldots, A_t$ are readily found from equations (11.6).

Knowing the values of the A_i and B_i, we can split the rational function of (3) into partial fractions to get

$$\phi(\theta) = \frac{p_1}{1 - q_1\theta} + \frac{p_2}{1 - q_2\theta} + \cdots + \frac{p_t}{1 - q_t\theta}.$$

Comparing this with (1), we see that

$$\begin{aligned}
m_1 &= p_1 & X_1 &= q_1 \\
m_2 &= p_2 & X_2 &= q_2 \\
&\;\;\vdots & &\;\;\vdots \\
m_t &= p_t & X_t &= q_t
\end{aligned}$$

and the system (11.5) is solved.

Remark 11.8 The polynomial

$$\sigma(\theta) = 1 + B_1\theta + B_2\theta^2 + \cdots + B_t\theta^t$$
$$= (1 - X_1\theta)(1 - X_2\theta) \cdots (1 - X_t\theta)$$

is what coding theorists call the *error-locator polynomial*; its zeros are the inverses of the error locations $X_1, X_2, \ldots, X_t$. The polynomial

$$\omega(\theta) = A_1 + A_2\theta + \cdots + A_t\theta^{t-1}$$

is what coding theorists call the *error-evaluator polynomial*. Once we have found the error locations, we can use the evaluator polynomial to calculate the error magnitudes.

Let us illustrate the above method by an example.

Example 11.9 Consider the 3-error-correcting code over $GF(11)$ with parity-check matrix

$$\begin{bmatrix} 1 & 1 & 1 & \cdots & 1 \\ 1 & 2 & 3 & \cdots & 10 \\ 1 & 2^2 & 3^2 & \cdots & 10^2 \\ 1 & 2^3 & 3^3 & \cdots & 10^3 \\ 1 & 2^4 & 3^4 & \cdots & 10^4 \\ 1 & 2^5 & 3^5 & \cdots & 10^5 \end{bmatrix}.$$

Suppose we have received a vector whose syndrome has been calculated to be

$$(S_1, S_2, S_3, S_4, S_5, S_6) = (2, 8, 4, 5, 3, 2).$$

Assuming at most 3 errors, in positions X_1, X_2, X_3 of respective magnitudes m_1, m_2, m_3 we have

$$\phi(\theta) = \frac{m_1}{1 - X_1\theta} + \frac{m_2}{1 - X_2\theta} + \frac{m_3}{1 - X_3\theta} = \frac{A_1 + A_2\theta + A_3\theta^2}{1 + B_1\theta + B_2\theta^2 + B_3\theta^3},$$

where, by 11.6 and 11.7, the A_i and B_i satisfy

$$A_1 = 2$$
$$A_2 = 8 + 2B_1$$
$$A_3 = 4 + 8B_1 + 2B_2$$
$$0 = 5 + 4B_1 + 8B_2 + 2B_3$$
$$0 = 3 + 5B_1 + 4B_2 + 8B_3$$
$$0 = 2 + 3B_1 + 5B_2 + 4B_3.$$

Solving first the last three equations for B_1, B_2 and B_3 gives

$B_1 = 5$, $B_2 = 10$, $B_3 = 8$, $A_1 = 2$, $A_2 = 7$ and $A_3 = 9$. Therefore

$$\phi(\theta) = \frac{2 + 7\theta + 9\theta^2}{1 + 5\theta + 10\theta^2 + 8\theta^3}.$$

To split this into partial fractions we must factorize the denominator. Because there is only a finite number of field elements, the simplest way to find the zeros of the denominator is by trial and error. In this case we find that the zeros are 4, 5 and 9. The error positions are the inverses of these values, i.e. 3, 9, and 5, and we now have

$$\phi(\theta) = \frac{2 + 7\theta + 9\theta^2}{(1 - 3\theta)(1 - 5\theta)(1 - 9\theta)} = \frac{m_1}{1 - 3\theta} + \frac{m_2}{1 - 5\theta} + \frac{m_3}{1 - 9\theta}. \quad (1)$$

Now m_1 is given by multiplying through by $1 - 3\theta$ and then putting $3\theta = 1$, i.e. $\theta = 3^{-1} = 4$, to get

$$m_1 = \frac{2 + 7 \cdot 4 + 9 \cdot 4^2}{(1 - 5 \cdot 4)(1 - 9 \cdot 4)} = 4.$$

The reader familiar with partial fractions may recognize this method as a 'cover-up' rule. Similarly, m_2 is obtained from the left-hand side of (1) by 'covering up' the factor $1 - 5\theta$ and putting $\theta = 5^{-1} = 9$. This gives $m_2 = 2$ and similarly we get $m_3 = 7$. Thus the error vector is

$$0 \ 0 \ 4 \ 0 \ 2 \ 0 \ 0 \ 0 \ 7 \ 0.$$

Notes (1) If the number of errors which actually have occurred is e, where $e < t$, then $m_{e+1} = m_{e+2} = \cdots = m_t = 0$ so that $\phi(\theta)$ becomes

$$\frac{A_1 + A_2\theta + \cdots + A_e\theta^{e-1}}{1 + B_1\theta + \cdots + B_e\theta^e}.$$

We therefore require a solution of equations (11.7) for which

$$B_{e+1} = B_{e+2} = \cdots = B_t = 0.$$

It will not be obvious from the received vector, nor from the syndrome, what the number e of errors is, but if $e < t$, then only the first e equations of (11.7) will be linearly independent, the remaining $t - e$ equations being dependent on these. So when solving the system (11.7) we must find the maximum number e of linearly independent equations and put $B_{e+1} = B_{e+2} = \cdots = B_t = 0$.

For example, suppose in Example 11.9 that the syndrome has been found to be $(5, 6, 0, 3, 7, 5)$. Then equations (11.7) become

$$6B_2 + 5B_3 = 8$$
$$3B_1 \qquad + 6B_3 = 4$$
$$7B_1 + 3B_2 \qquad = 6.$$

Eliminating B_1 from the last two equations gives

$$3B_2 + 8B_3 = 4$$

which is just a scalar multiple of the first. So we put $B_3 = 0$ and solve the first two equations for B_1 and B_2 to get $B_1 = 5$ and $B_2 = 5$. We then have $A_1 = 5$ and $A_2 = 9$. So

$$\phi(\theta) = \frac{5 + 9\theta}{1 + 5\theta + 5\theta^2},$$

which gives, on splitting into partial fractions,

$$\frac{2}{1 - \theta} + \frac{3}{1 - 5\theta}.$$

Thus we assume that there are just two errors, in position 1 of magnitude 2, and in position 5 of magnitude 3.

(2) When the error-locator and error-evaluator polynomials $\sigma(\theta)$ and $\omega(\theta)$ (defined in Remark 11.8) have been found, and the error locations $X_1, X_2, \ldots, X_e$ determined, then, as we saw in Example 11.9, the error magnitudes are given by

$$m_j = \frac{\omega(X_j^{-1})}{\displaystyle\prod_{\substack{i=1 \\ i \neq j}}^{e} (1 - X_i X_j^{-1})} \qquad \text{for } j = 1, 2, \ldots, e. \qquad (11.10)$$

This is why $\omega(\theta)$ is called the error-evaluator polynomial. We now summarize the general algorithm.

Outline of the error-correction procedure (assuming $\leqslant t$ errors)

Step 1 Calculate the syndrome $(S_1, S_2, \ldots, S_{2t})$ of the received vector.

Step 2 Determine the maximum number of equations in system (11.7) which are linearly independent. This is the number e of errors which actually occurred.

Step 3 Set $B_{e+1}, B_{e+2}, \ldots, B_t$ all equal to zero and solve the first e equations of (11.7) for $B_1, B_2, \ldots, B_e$.

Step 4 Find the zeros of the error locator polynomial

$$1 + B_1\theta + B_2\theta^2 + \cdots + B_e\theta^e$$

by substituting each of the non-zero elements of $GF(q)$.

Step 5 Find $A_1, A_2, \ldots, A_e$ from system (11.6) and find each error magnitude m_j by substituting X_j^{-1} in the error-evaluator polynomial $A_1 + A_2\theta + \cdots + A_e\theta^{e-1}$ and dividing by the product of the factors $1 - X_iX_j^{-1}$ for $i = 1, 2, \ldots, e$ with $i \neq j$.

Notes (1) If in Step 3 we solve the system (11.7) by reducing to upper triangular form, then we can automatically carry out Step 2 at the same time.

(2) The above procedure is essentially that used by coding theorists today, although Ramanujan's consideration of partial fractions is not used explicitly.

(3) The computations involved in the above scheme may all be performed very quickly with the exception of Step 3, in which we are required to solve the matrix equation

$$\begin{bmatrix} S_1 & S_2 & S_3 & \cdots & S_e \\ S_2 & S_3 & S_4 & \cdots & S_{e+1} \\ S_3 & & & & \\ \vdots & & & & \vdots \\ S_e & S_{e+1} & S_{e+2} & \cdots & S_{2e-1} \end{bmatrix} \begin{bmatrix} B_e \\ B_{e-1} \\ \vdots \\ B_1 \end{bmatrix} = \begin{bmatrix} -S_{e+1} \\ -S_{e+2} \\ \vdots \\ -S_{2e} \end{bmatrix}.$$

For example, if we were to solve the system by inverting the $e \times e$ matrix, then the number of computations needed would be proportional to e^3. This might be reasonable for small t, but if we need to correct a large number of errors we require a more efficient method of solution. Various refinements have been found which greatly reduce the amount and complexity of computation.

Note that the $e \times e$ matrix above is not arbitrary in form, but has the property known as 'persymmetry'; that is, the entries in any diagonal perpendicular to the main diagonal are all identical.

Berlekamp (1968) and Massey (1969) were able to use this additional structure to obtain a method of solving the equations in a computationally much simpler way. This involved converting the problem to one involving linear-feedback shift registers; details may be found in Peterson and Weldon (1972), MacWilliams and Sloane (1977) or Blahut (1983). An alternative algorithm (see same references) involves the clever use of the Euclidean algorithm for polynomials. This algorithm is perhaps easier to understand than the Berlekamp–Massey algorithm, though it is thought to be less efficient in practice.

(4)   Since we require that $n \leqslant q - 1$ in constructing the above codes, it may look as though the methods of this chapter have no applicability to binary codes. However, binary BCH codes indeed exist and are extremely important. A binary BCH code may be defined by constructing a certain matrix whose entries belong to a field of order 2^h and then converting this to a parity-check matrix for a binary code by identifying each element of $GF(2^h)$ with a binary h-tuple (written as a column vector) in a natural way. These BCH codes are discussed extensively in several of the standard texts on coding theory. It is hoped that for the reader who wishes to study BCH codes further, the above treatment will facilitate his understanding of the more general case.

Concluding remarks

(1)   Apart from the ISBN code, modulus 11 decimal codes are now widely used, mainly for error detection rather than correction. One of the earliest uses was in the allocation of registration numbers to the entire population of Norway in a scheme devised by Selmer (cf. 1967). Selmer's code, defined in Exercise 11.6, satisfies two parity-check equations and is designed to detect all single errors and various types of commonly occurring multiple errors. Before devising his code, in order to ascertain which psychological errors occurred most frequently, Selmer analysed the census returns of 1960 for the population of Oslo. In this census, the public had filled in the date of birth themselves, and comparison of these entries with those in the public register had revealed about 8000 inconsistencies, which were on record in Oslo. Selmer actually received only 7000 of these; the remaining

thousand were people who had also written their name incorrectly and so belonged to another file!

(2) For a survey of various types of error-detecting decimal codes, see Verhoeff (1969). This includes, in Chapter 5, the first example of a *pure* decimal code which detects all single errors and all transpositions.

(3) In 1970, Goppa discovered codes which are an important generalization of BCH codes, and whose decoding can be carried out in essentially the same way. McEliece (1977) asserts that 'it is fairly clear that the deepest and most impressive result in coding theory is the algebraic decoding of BCH–Goppa codes'. It has been the aim of this chapter to give the essential flavour of this result assuming nothing more than standard results from first-year undergraduate mathematics.

Exercises 11

11.1 Using the code of Example 11.3, decode the received vector 1204000910.

11.2 Find a generator matrix for the $[10, 6]$-code of Example 11.3.

11.3 For the code of Example 11.9, find the error vectors corresponding to the syndromes

$$(1, 7, 5, 2, 3, 10) \quad \text{and} \quad (9, 7, 7, 10, 8, 3).$$

11.4 Suppose we wished to give each person in a population of some 200 000 a personal identity codeword composed of letters of the English alphabet. Devise a suitable code of reasonably short length which is double-error-correcting.

11.5 When decoding a BCH code of minimum distance $2t + 1$, suppose the error locations are found to be $X_1, X_2, \ldots, X_e$. Show that the error magnitude m_j in position X_j is given by

$$m_j = -X_j \omega(X_j^{-1}) / \sigma'(X_j^{-1}),$$

where $\omega(\theta)$ is the error-evaluator polynomial and $\sigma'(\theta)$ denotes the derivative of the error-locator polynomial $\sigma(\theta)$.

11.6 Every person in Norway has an 11-digit decimal registration number $x_1 x_2 \cdots x_{11}$, where $x_1 x_2 \cdots x_6$ is the date of birth, $x_7x_8x_9$ is a personal number and x_{10} and x_{11} are check digits defined by

$$x_{10} \equiv -(2x_9 + 5x_8 + 4x_7 + 9x_6 + 8x_5 + x_4 + 6x_3 + 7x_2 + 3x_1)$$

$$(\text{mod } 11)$$

and

$$x_{11} \equiv -(2x_{10} + 3x_9 + 4x_8 + 5x_7 + 6x_6 + 7x_5$$
$$+ 2x_4 + 3x_3 + 4x_2 + 5x_1)\,(\text{mod } 11).$$

Write down a parity-check matrix for the code (regarded as a code over $GF(11)$). If the code is used only for error detection, will all double errors be detected? If not, which double errors will fail to be detected?

12 Cyclic codes

Cyclic codes form an important class of codes for several reasons. From a theoretical point of view they possess a rich algebraic structure, while practically they can be efficiently implemented by means of simple devices known as shift registers. Furthermore, many important codes, such as binary Hamming codes, Golay codes and BCH codes, are equivalent to cyclic codes.

Definition A code C is *cyclic* if (i) C is a linear code and (ii) any cyclic shift of a codeword is also a codeword, i.e. whenever $a_0 a_1 \cdots a_{n-1}$ is in C, then so is $a_{n-1} a_0 a_1 \cdots a_{n-2}$.

Examples 12.1 (i) The binary code $\{000, 101, 011, 110\}$ is cyclic.

(ii) The code of Example 2.23, which we now know as the Hamming code Ham $(3, 2)$, is cyclic. (Note that each codeword of the form $\mathbf{a}_i$ is the first cyclic shift of its predecessor and so is each $\mathbf{b}_i$.)

(iii) The binary linear code $\{0000, 1001, 0110, 1111\}$ is not cyclic, but it is *equivalent* to a cyclic code; interchanging the third and fourth coordinates gives the cyclic code $\{0000, 1010, 0101, 1111\}$.

(iv) Consider the ternary Hamming code Ham $(2, 3)$ with generator matrix $\begin{bmatrix} 1 & 0 & 1 & 1 \\ 0 & 1 & 1 & 2 \end{bmatrix}$. From the list of codewords found in Exercise 5.7, we see that the code is not cyclic. But is Ham $(2, 3)$ equivalent to a cyclic code? The answer will be given in Example 12.13 (see also Exercise 12.22).

When considering cyclic codes we number the coordinate positions $0, 1, \ldots, n-1$. This is because it is useful to let a vector $a_0 a_1 \cdots a_{n-1}$ in $V(n, q)$ correspond to the polynomial $a_0 + a_1 x + \cdots + a_{n-1} x^{n-1}$.

Polynomials

From now on we will denote the field $GF(q)$ by F_q, or simply by F (with q understood). We denote by $F[x]$ the set of polynomials in x with coefficients in F. If $f(x) = f_0 + f_1 x + \cdots + f_m x^m$ is a polynomial with $f_m \neq 0$, then m is called the *degree* of $f(x)$, denoted $\deg f(x)$. (By convention the degree of the zero polynomial is $-\infty$.) The coefficient f_m is then called the *leading coefficient*. A polynomial is called *monic* if its leading coefficient is 1.

Polynomials in $F[x]$ can be added, subtracted and multiplied in the usual way. $F[x]$ is an example of an algebraic structure known as a *ring*, for it satisfies the first seven of the eight field axioms (see Chapter 3). Note that $F[x]$ is not a field since polynomials of degree greater than zero do not have multiplicative inverses. Observe also that if $f(x)$, $g(x) \in F[x]$, then $\deg (f(x)g(x)) = \deg f(x) + \deg g(x)$.

The division algorithm for polynomials

The division algorithm states that, for every pair of polynomials $a(x)$ and $b(x) \neq 0$ in $F[x]$, there exists a unique pair of polynomials $q(x)$, the quotient, and $r(x)$, the remainder, such that

$$a(x) = q(x)b(x) + r(x),$$

where $\deg r(x) < \deg b(x)$.

This is analogous to the familiar division algorithm for the ring $\mathbb{Z}$ of integers. The polynomials $q(x)$ and $r(x)$ can be obtained by ordinary long division of polynomials.

For example, in $F_2[x]$, we can divide $x^3 + x + 1$ by $x^2 + x + 1$ as follows.

$$
\begin{array}{r}
x + 1 \\
x^2 + x + 1 \,\overline{)\; x^3 \qquad + x + 1} \\
\underline{x^3 + x^2 + x \qquad\quad} \\
x^2 \qquad + 1 \\
\underline{x^2 + x + 1} \\
x
\end{array}
$$

Hence $x^3 + x + 1 = (x + 1)(x^2 + x + 1) + x$ is the desired expression of $x^3 + x + 1$ as $q(x)(x^2 + x + 1) + r(x)$.

The ring of polynomials modulo f(x)

The ring $F[x]$ of polynomials over F is analogous in many ways to the ring $\mathbb{Z}$ of integers. Just as we can consider integers modulo some fixed integer m to get the ring Z_m (see Chapter 3), we can consider polynomials in $F[x]$ modulo some fixed polynomial $f(x)$.

Let $f(x)$ be a fixed polynomial in $F[x]$. Two polynomials $g(x)$ and $h(x)$ in $F[x]$ are said to be *congruent modulo $f(x)$*, symbolized by

$$g(x) \equiv h(x) \,(\mathrm{mod}\, f(x)),$$

if $g(x) - h(x)$ is divisible by $f(x)$.

By the division algorithm, any polynomial $a(x)$ in $F[x]$ is congruent modulo $f(x)$ to a unique polynomial $r(x)$ of degree less than $\deg f(x)$; $r(x)$ is just the principal remainder when $a(x)$ is divided by $f(x)$.

We denote by $F[x]/f(x)$ the set of polynomials in $F[x]$ of degree less than $\deg f(x)$, with addition and multiplication carried out modulo $f(x)$ as follows.

Suppose $a(x)$ and $b(x)$ belong to $F[x]/f(x)$. Then the sum $a(x) + b(x)$ in $F[x]/f(x)$ is the same as the sum in $F[x]$, because $\deg (a(x) + b(x)) < \deg f(x)$. The product $a(x)b(x)$ in $F[x]/f(x)$ is the unique polynomial of degree less than $\deg f(x)$ to which $a(x)b(x)$ (as a product in $F[x]$) is congruent modulo $f(x)$.

For example, let us calculate $(x + 1)^2$ in $F_2[x]/(x^2 + x + 1)$. We have

$$(x + 1)^2 = x^2 + 2x + 1 = x^2 + 1 \equiv x \,(\mathrm{mod}\, x^2 + x + 1).$$

Thus $(x + 1)^2 = x$ in $F_2[x]/(x^2 + x + 1)$.

Just as Z_m is a ring, so also is $F[x]/f(x)$; it is called the *ring of polynomials (over F) modulo $f(x)$*.

If $f(x) \in F_q[x]$ has degree n, then the ring $F_q[x]/f(x)$ consists of polynomials of degree $\leq n - 1$. Each of the n coefficients of such a polynomial belongs to F_q and so

$$|F_q[x]/f(x)| = q^n.$$

Example 12.2 The addition and multiplication tables for $F_2[x]/$

$(x^2 + x + 1)$ are easily found to be:

+	0	1	x	1+x		·	0	1	x	1+x
0	0	1	x	1+x		0	0	0	0	0
1	1	0	1+x	x		1	0	1	x	1+x
x	x	1+x	0	1		x	0	x	1+x	1
1+x	1+x	x	1	0		1+x	0	1+x	1	x

We see that this is more than just a ring. Every non-zero element has a multiplicative inverse and so $F_2[x]/(x^2 + x + 1)$ is actually a field. In fact, we have precisely the field of order 4 given in Example 3.6(3), with x and $1 + x$ corresponding to a and b respectively.

It is certainly not the case that $F[x]/f(x)$ is a field for any choice of $f(x)$; consider, for example, the multiplication table of $F_2[x]/(x^2 + 1)$ (see Exercise 12.2). The special property of $f(x)$ which makes $F[x]/f(x)$ a field is that of being 'irreducible', which we now define.

Definition   A polynomial $f(x)$ in $F[x]$ is said to be *reducible* if $f(x) = a(x)b(x)$, where $a(x), b(x) \in F[x]$ and $\deg a(x)$ and $\deg b(x)$ are both smaller than $\deg f(x)$. If $f(x)$ is not reducible, it is called *irreducible*.

Just as any positive integer can be factorized uniquely into a product of prime numbers, any monic polynomial in $F[x]$ can be factorized uniquely into a product of irreducible monic polynomials.

The following simple observations are often useful when factorizing a polynomial.

Lemma 12.3
  (i)   A polynomial $f(x)$ has a linear factor $x - a$ if and only if $f(a) = 0$.
  (ii) A polynomial $f(x)$ in $F[x]$ of degree 2 or 3 is irreducible if and only if $f(a) \neq 0$ for all a in F.
  (iii) Over any field, $x^n - 1 = (x - 1)(x^{n-1} + x^{n-2} + \cdots + x + 1)$ (the second factor may well be further reducible).

Proof   (i)   If $f(x) = (x - a)g(x)$, then certainly $f(a) = 0$. On the other hand, suppose $f(a) = 0$. By the division algorithm, $f(x) =$

$q(x)(x-a)+r(x)$, where $\deg r(x)<1$. So $r(x)$ is a constant, which must be zero since $0=f(a)=r(a)$.

(ii) A polynomial of degree 2 or 3 is reducible if and only if it has at least one linear factor. The result is now immediate from (i).

(iii) By (i), $x-1$ is a factor of x^n-1 and long division of x^n-1 by $x-1$ gives the other factor.

Example 12.4 (i) Factorize x^3-1 in $F_2[x]$ into irreducible polynomials.

(ii) Factorize x^3-1 in $F_3[x]$ into irreducible polynomials.

Solution By 12.3(iii), $x^3-1=(x-1)(x^2+x+1)$ over any field.

(i) By 12.3(ii), x^2+x+1 is irreducible in $F_2[x]$.

(ii) By 12.3(i), in $F_3[x]$, $x-1$ is a factor of x^2+x+1, and we get the factorization $x^3-1=(x-1)^3$.

The finite fields $GF(p^h)$, $h>1$

The property in $F[x]$ of a polynomial being irreducible corresponds exactly to the property in $\mathbb{Z}$ of a number being prime. We showed in Theorem 3.5 that the ring Z_m is a field if and only if m is prime and the following may be proved in exactly the same way.

Theorem 12.5 The ring $F[x]/f(x)$ is a field if and only if $f(x)$ is irreducible in $F[x]$.

Proof This is left to Exercise 12.3.

Although we do not show it here, it can be shown that for any prime number p and for any positive integer h, there *exists* an irreducible polynomial over $GF(p)$ of degree h. This result, together with Theorem 12.5, gives the existence of the fields $GF(p^h)$ for all integers $h\geq1$. As we remarked in Theorem 3.2, these are essentially the only finite fields.

Back to cyclic codes

Returning from our excursion to look at fields of general order, we now fix $f(x)=x^n-1$ for the remainder of the chapter, for we shall soon see that the ring $F[x]/(x^n - 1)$ of polynomials modulo $x^n - 1$ is the natural one to consider in the context of cyclic codes. For simplicity we shall write $F[x]/(x^n - 1)$ as R_n, where the field $F = F_q$ will be understood.

Since $x^n \equiv 1 \pmod{x^n - 1}$, we can reduce any polynomial modulo $x^n - 1$ simply by replacing x^n by 1, x^{n+1} by x, x^{n+2} by x^2 and so on. There is no need to write out long divisions by $x^n - 1$.

Let us now identify a vector $a_0 a_1 \cdots a_{n-1}$ in $V(n, q)$ with the polynomial

$$a(x) = a_0 + a_1 x + \cdots + a_{n-1} x^{n-1}$$

in R_n. We shall simultaneously view a code as a subset of $V(n, q)$ and as a subset of R_n. Note that addition of vectors and multiplication of a vector by a scalar in R_n corresponds exactly to those operations in $V(n, q)$. Now consider what happens when we multiply the polynomial $a(x)$ by x. In R_n, we have

$$x \cdot a(x) = a_0 x + a_1 x^2 + \cdots + a_{n-1} x^n$$
$$= a_{n-1} + a_0 x + \cdots + a_{n-2} x^{n-1},$$

which is the vector $a_{n-1} a_0 \cdots a_{n-2}$. Thus *multiplying by x corresponds to performing a single cyclic shift*. Multiplying by x^m corresponds to a cyclic shift through m positions.

The following theorem gives the algebraic characterization of cyclic codes.

Theorem 12.6   A code C in R_n is a cyclic code if and only if C satisfies the following two conditions:
(i) $a(x), b(x) \in C \Rightarrow a(x) + b(x) \in C$,
(ii) $a(x) \in C$ and $r(x) \in R_n \Rightarrow r(x)a(x) \in C$.
[Note that (ii) does not just say that C must be closed under multiplication; it says that C must be closed under multiplication *by any element of R_n*. The reader who is familiar with ring theory will recognize that Theorem 12.6 says that cyclic codes are precisely the 'ideals' of the ring R_n.]

Proof   Suppose C is a cyclic code in R_n. Then C is linear and so (i) holds. Now suppose $a(x) \in C$ and $r(x) = r_0 + r_1 x + \cdots + r_{n-1} x^{n-1} \in R_n$. Since multiplication by x corresponds to a cyclic shift, we have $x \cdot a(x) \in C$ and then $x \cdot (xa(x)) = x^2 a(x) \in C$ and so on. Hence

$$r(x)a(x) = r_0 a(x) + r_1 x a(x) + \cdots + r_{n-1} x^{n-1} a(x)$$

is also in C since each summand is in C. Thus (ii) also holds.

Now suppose (i) and (ii) hold. Taking $r(x)$ to be a scalar, the conditions imply that C is linear. Taking $r(x) = x$ in (ii) shows that C is cyclic.

We now give an easy way of constructing examples of cyclic codes.

Let $f(x)$ be any polynomial in R_n and let $\langle f(x) \rangle$ denote the subset of R_n consisting of all multiples of $f(x)$ (reduced modulo $x^n - 1$), i.e.

$$\langle f(x) \rangle = \{r(x)f(x) \mid r(x) \in R_n\}.$$

Theorem 12.7 For any $f(x) \in R_n$, the set $\langle f(x) \rangle$ is a cyclic code; it is called the code *generated* by $f(x)$.

Proof We check conditions (i) and (ii) of Theorem 12.6.
   (i)   If $a(x)f(x)$ and $b(x)f(x) \in \langle f(x) \rangle$, then

$$a(x)f(x) + b(x)f(x) = (a(x) + b(x))f(x) \in \langle f(x) \rangle.$$

   (ii)   If $a(x)f(x) \in \langle f(x) \rangle$ and $r(x) \in R_n$, then

$$r(x)(a(x)f(x)) = (r(x)a(x))f(x) \in \langle f(x) \rangle.$$

Example 12.8 Consider the code $C = \langle 1 + x^2 \rangle$ in R_3 (with $F = GF(2)$). Multiplying $1 + x^2$ by each of the eight elements of R_3 (and reducing modulo $x^3 - 1$) produces only four distinct codewords, namely $0, 1 + x, 1 + x^2$ and $x + x^2$. Thus C is the code $\{000, 110, 101, 011\}$ of Example 12.1(i).

We next show that the above easy way of constructing cyclic codes is essentially the *only* way, i.e. any cyclic code can be generated by a polynomial. (In the terminology of ring theory, this says that every ideal in R_n is a 'principal ideal'.)

Theorem 12.9 Let C be a non-zero cyclic code in R_n. Then
   (i)   there exists a unique monic polynomial $g(x)$ of smallest degree in C,
   (ii) $C = \langle g(x) \rangle$,
   (iii) $g(x)$ is a factor of $x^n - 1$.

Proof (i) Suppose $g(x)$ and $h(x)$ are both monic polynomials in C of smallest degree. Then $g(x) - h(x) \in C$ and has smaller degree. This gives a contradiction if $g(x) \neq h(x)$, for then a suitable scalar multiple of $g(x) - h(x)$ is monic, is in C, and is of smaller degree then $\deg g(x)$.

(ii) Suppose $a(x) \in C$. By the division algorithm for $F[x]$, $a(x) = q(x)g(x) + r(x)$, where $\deg r(x) < \deg g(x)$. But $r(x) = a(x) - q(x)g(x) \in C$, by the properties of a cyclic code given in Theorem 12.6. By the minimality of $\deg g(x)$, we must have $r(x) = 0$ and so $a(x) \in \langle g(x) \rangle$.

(iii) By the division algorithm,

$$x^n - 1 = q(x)g(x) + r(x),$$

where $\deg r(x) < \deg g(x)$. But then $r(x) \equiv -q(x)g(x) \pmod{x^n - 1}$, and so $r(x) \in \langle g(x) \rangle$. By the minimality of $\deg g(x)$, we must have $r(x) = 0$, which implies that $g(x)$ is a factor of $x^n - 1$.

Definition In a non-zero cyclic code C the monic polynomial of least degree, given by Theorem 12.9, is called the *generator polynomial* of C.

Note that a cyclic code C may contain polynomials other than the generator polynomial which also generate C. For example, the code of Example 12.8 is generated by $1 + x^2$, but its generator polynomial is $1 + x$.

The third part of Theorem 12.9 gives a recipe for finding all cyclic codes of given length n. All we need is the factorization of $x^n - 1$ into irreducible monic polynomials.

Example 12.10 We will find all the binary cyclic codes of length 3. By Example 12.4(i), $x^3 - 1 = (x + 1)(x^2 + x + 1)$, where $x + 1$ and $x^2 + x + 1$ are irreducible over $GF(2)$. So, by Theorem 12.9, the following is a complete list of binary cyclic codes of length 3.

Generator polynomial	Code in R_3	Corresponding Code in $V(3, 2)$
1	all of R_3	all of $V(3, 2)$
$x + 1$	$\{0, 1 + x, x + x^2, 1 + x^2\}$	$\{000, 110, 011, 101\}$
$x^2 + x + 1$	$\{0, 1 + x + x^2\}$	$\{000, 111\}$
$x^3 - 1 = 0$	$\{0\}$	$\{000\}$

Lemma 12.11 Let $g(x) = g_0 + g_1 x + \cdots + g_r x^r$ be the generator polynomial of a cyclic code. Then g_0 is non-zero.

Proof Suppose $g_0 = 0$. Then $x^{n-1} g(x) = x^{-1} g(x)$ is a codeword of C of degree $r - 1$, contradicting the minimality of $\deg g(x)$.

By definition, a cyclic code is linear. It would be handy if immediately from the generator polynomial $g(x)$ we could deduce the dimension of the code and also write down a generator matrix. The next theorem shows that we can do both.

Theorem 12.12 Suppose C is a cyclic code with generator polynomial

$$g(x) = g_0 + g_1 x + \cdots + g_r x^r$$

of degree r. Then $\dim (C) = n - r$ and a generator matrix for C is

$$G = \begin{bmatrix} g_0 & g_1 & g_2 & \cdots & & g_r & 0 & 0 & \cdots & 0 \\ 0 & g_0 & g_1 & g_2 & \cdots & & g_r & 0 & \cdots & 0 \\ 0 & 0 & g_0 & g_1 & g_2 & \cdots & & g_r & & \vdots \\ \vdots & \vdots & & & \ddots & & & & \ddots & 0 \\ 0 & 0 & \cdots & 0 & g_0 & g_1 & g_2 & \cdots & & g_r \end{bmatrix}$$

Proof The $n - r$ rows of the above matrix G are certainly linearly independent because of the echelon of non-zero g_0s with 0s below. These $n - r$ rows represent the codewords $g(x)$, $xg(x)$, $x^2 g(x), \ldots, x^{n-r-1} g(x)$, and it remains only to show that every codeword in C can be expressed as a linear combination of them. The proof of Theorem 12.9(ii) shows that if $a(x)$ is a codeword of C, then

$$a(x) = q(x) g(x)$$

for some polynomial $q(x)$, and that this is an equality of polynomials within $F[x]$, not requiring any reduction modulo $x^n - 1$. Since $\deg a(x) < n$, it follows that $\deg q(x) < n - r$. Hence

$$q(x) g(x) = (q_0 + q_1 x + \cdots + q_{n-r-1} x^{n-r-1}) g(x)$$
$$= q_0 g(x) + q_1 x g(x) + \cdots + q_{n-r-1} x^{n-r-1} g(x),$$

which is the desired linear combination.

Example 12.13 Find all the ternary cyclic codes of length 4 and write down a generator matrix for each of them.

Solution Over $GF(3)$, the factorization of $x^4 - 1$ into irreducible polynomials is

$$x^4 - 1 = (x - 1)(x^3 + x^2 + x + 1) = (x - 1)(x + 1)(x^2 + 1).$$

So there are $2^3 = 8$ divisors of $x^4 - 1$ in $F_3[x]$, each of which generates a cyclic code. By Theorem 12.9, these are the only ternary cyclic codes of length 4. The codes are specified below by their generator polynomials, and the corresponding generator matrices are given by Theorem 12.12. Note that neither of the two-dimensional codes has minimum distance 3 and so the ternary Hamming $[4, 2, 3]$-code is not cyclic, thus answering the question posed in Example 12.1(iv).

Generator polynomial	Generator matrix
1	$[I_4]$
$x - 1$	$\begin{bmatrix} -1 & 1 & 0 & 0 \\ 0 & -1 & 1 & 0 \\ 0 & 0 & -1 & 1 \end{bmatrix}$
$x + 1$	$\begin{bmatrix} 1 & 1 & 0 & 0 \\ 0 & 1 & 1 & 0 \\ 0 & 0 & 1 & 1 \end{bmatrix}$
$x^2 + 1$	$\begin{bmatrix} 1 & 0 & 1 & 0 \\ 0 & 1 & 0 & 1 \end{bmatrix}$
$(x - 1)(x + 1) = x^2 - 1$	$\begin{bmatrix} -1 & 0 & 1 & 0 \\ 0 & -1 & 0 & 1 \end{bmatrix}$
$(x - 1)(x^2 + 1) = x^3 - x^2 + x - 1$	$[-1 \quad 1 \; -1 \quad 1]$
$(x + 1)(x^2 + 1) = x^3 + x^2 + x + 1$	$[\;1 \quad 1 \quad 1 \quad 1]$
$x^4 - 1 = 0$	$[\;0 \quad 0 \quad 0 \quad 0]$

The check polynomial and the parity-check matrix of a cyclic code

The generator matrix of a cyclic code as given by Theorem 12.12 is not in standard form. Our usual method of writing down a parity-check matrix from the standard form of G (via Theorem 7.6) is therefore not appropriate for cyclic codes. However, there is a natural choice of parity-check matrix for a cyclic code. This is closely related to the so-called 'check polynomial', which we define first.

Let C be a cyclic $[n, k]$-code with generator polynomial $g(x)$. By Theorem 12.9, $g(x)$ is a factor of $x^n - 1$ and so

$$x^n - 1 = g(x)h(x),$$

for some polynomial $h(x)$. Since $g(x)$ is monic, so also is $h(x)$. By Theorem 12.12, $g(x)$ has degree $n - k$ and so $h(x)$ has degree k. This polynomial $h(x)$ is called the *check polynomial* of C. The reason for this name is apparent from the following theorem.

Theorem 12.14 Suppose C is a cyclic code in R_n with generator polynomial $g(x)$ and check polynomial $h(x)$. Then an element $c(x)$ of R_n is a codeword of C if and only if $c(x)h(x) = 0$.

Proof First note that, in R_n, $g(x)h(x) = x^n - 1 = 0$.
Hence $c(x) \in C \Rightarrow c(x) = a(x)g(x)$, for some $a(x) \in R_n$,

$$\Rightarrow c(x)h(x) = a(x)g(x)h(x)$$
$$= a(x) \cdot 0$$
$$= 0.$$

On the other hand, suppose $c(x)$ satisfies $c(x)h(x) = 0$. By the division algorithm, $c(x) = q(x)g(x) + r(x)$, where $\deg r(x) < n - k$. Then $c(x)h(x) = 0$ implies that $r(x)h(x) = 0$, i.e. $r(x)h(x) \equiv 0 \pmod{x^n - 1}$. But $\deg (r(x)h(x)) < n - k + k = n$, and so $r(x)h(x) = 0$ in $F[x]$. Hence $r(x) = 0$, and then $c(x) = q(x)g(x) \in C$.

In view of Theorem 12.14 and the fact that $\dim (\langle h(x) \rangle) = n - k = \dim (C^\perp)$, we might easily be fooled into thinking that $h(x)$ generates the dual code $C^\perp$. In general this is not so. The point is that the product of $c(x)$ and $h(x)$ being zero in R_n is not the same thing as the corresponding vectors in $V(n, q)$ being orthogonal. In the next theorem, however, we see that the condition $c(x)h(x) = 0$ in R_n does imply some useful orthogonality relations which lead to a natural choice of parity-check matrix.

Theorem 12.15 Suppose C is a cyclic $[n, k]$-code with check polynomial

$$h(x) = h_0 + h_1x + \cdots + h_kx^k.$$

Then

  (i)   a parity-check matrix for C is

$$H = \begin{bmatrix} h_k & h_{k-1} & \cdots & h_0 & 0 & 0 & \cdots & 0 \\ 0 & h_k & h_{k-1} & \cdots & h_0 & 0 & \cdots & 0 \\ & \cdot & \cdot & & & \cdot & & \vdots \\ & & \cdot & \cdot & & & \cdot & 0 \\ 0 & \cdots & 0 & h_k & h_{k-1} & \cdots & & h_0 \end{bmatrix}$$

  (ii) $C^\perp$ is a cyclic code generated by the polynomial

$$\bar{h}(x) = h_k + h_{k-1}x + \cdots + h_0x^k.$$

Proof (i) By Theorem 12.14, a polynomial $c(x) = c_0 + c_1x + \cdots + c_{n-1}x^{n-1}$ is a codeword if and only if $c(x)h(x) = 0$. Now for $c(x)h(x)$ to be zero, then in particular the coefficients of $x^k, x^{k+1}, \ldots, x^{n-1}$ must all be zero, i.e.

$$\begin{aligned} c_0h_k + c_1h_{k-1} + \cdots &+ c_kh_0 &= 0 \\ c_1h_k &+ c_2h_{k-1} + \cdots + c_{k+1}h_0 &= 0 \\ &\ddots &\vdots \\ c_{n-k-1}h_k + \cdots &+ c_{n-1}h_0 &= 0. \end{aligned}$$

Thus any codeword $c_0c_1 \cdots c_{n-1}$ of C is orthogonal to the vector $h_kh_{k-1} \cdots h_000 \cdots 0$ and to its cyclic shifts. So the rows of the matrix H given in the statement of the theorem are all codewords of $C^\perp$. We have already observed that $h(x)$ is monic of degree k and so $h_k = 1$; thus the echelon of 1s with zeros below in H ensures that the rows of H are linearly independent. The number of rows of H is $n - k$, which is the dimension of $C^\perp$. Hence H is a generator matrix of $C^\perp$, i.e. a parity-check matrix for C.

  (ii) If we can show that $\bar{h}(x)$ is a factor of $x^n - 1$, then it will follow from Theorem 12.12 that $\langle \bar{h}(x) \rangle$ is a cyclic code whose generator matrix is the above matrix H, and hence that $\langle \bar{h}(x) \rangle = C^\perp$. We observe that $\bar{h}(x) = x^kh(x^{-1})$. Since $h(x^{-1})g(x^{-1}) = (x^{-1})^n - 1$, we have $x^kh(x^{-1})x^{n-k}g(x^{-1}) = x^n(x^{-n} - 1) = 1 - x^n$, and so $\bar{h}(x)$ is indeed a factor of $x^n - 1$.

Remarks (i) The polynomial $\bar{h}(x) = x^k h(x^{-1}) = h_k + h_{k-1}x + \cdots + h_0 x^k$ is called the *reciprocal polynomial* of $h(x)$; its coefficients are those of $h(x)$ *in reverse order*.

(ii) We may regard $\bar{h}(x)$ as the generator polynomial of $C^\perp$, though strictly speaking, in the non-binary case, one ought to multiply it by the scalar h_0^{-1} to make it monic.

(iii) The polynomial $h(x^{-1}) = x^{n-k}\bar{h}(x)$ is a member of $C^\perp$.

We have not yet discussed the minimum distance of cyclic codes. There are some classes of cyclic codes for which useful lower bounds on the minimum distance are known. For example, cyclic BCH codes can be constructed to have 'designed minimum distance' while there are codes called quadratic residue codes which satisfy a 'square root bound'. These codes and bounds are well treated in several of the more advanced texts. We concentrate here on finding the minimum distances of two particularly interesting cyclic codes, namely the two Golay codes. Our methods, while aimed directly at the codes in hand, nevertheless provide some insights into the more general methods.

The binary Golay code

In Chapter 9, we proved the existence of a perfect binary $[23, 12, 7]$-code G_{23} by exhibiting a generator matrix. We now show that this Golay code can be constructed in a more natural way as a cyclic code. The only knowledge we shall assume in advance is the factorization of $x^{23} - 1$ over $GF(2)$. [There is a clever method of finding the factors of $x^n - 1$ over $GF(q)$ in general (see, for example, Chapter 7, §5, of MacWilliams and Sloane (1977)) but we shall not dwell on this here. Alternatively one may find the factors by consulting tables (see, e.g., the same reference for a list of factors of $x^n - 1$ over $GF(2)$ for $n \le 63$).]

We begin then with the factorization

$$x^{23} - 1 = (x - 1)(x^{11} + x^{10} + x^6 + x^5 + x^4 + x^2 + 1)$$
$$\times (x^{11} + x^9 + x^7 + x^6 + x^5 + x + 1)$$
$$= (x - 1)g_1(x)g_2(x), \text{ say.}$$

Let C_1 be the code $\langle g_1(x) \rangle$ and let C_2 be the code $\langle g_2(x) \rangle$. By Theorem 12.12, C_1 is a $[23, 12]$-code. The object of the next few pages is to show that the minimum distance of C_1 is 7.

We observe that the polynomials $g_1(x)$ and $g_2(x)$ are reciprocals of each other, and so C_2 is equivalent to C_1. Remarkably, the knowledge that $x^{23} - 1 = (x - 1)g_1(x)\bar{g}_1(x)$, where $\bar{g}_1(x)$ denotes the reciprocal of $g_1(x)$, is all we need to show that $d(C_1) = 7$; we do not actually need to know what $g_1(x)$ is.

Remark 12.16 Although we do not show it here, $x^p - 1$ has a factorization over $GF(2)$ of the form $(x - 1)g_1(x)g_2(x)$, where $\langle g_1(x) \rangle$ and $\langle g_2(x) \rangle$ are equivalent codes, whenever p is a prime number of the form $8m \pm 1$. If p is of the form $8m - 1$ we also have $g_2(x) = \bar{g}_1(x)$. For example,

$$x^7 - 1 = (x - 1)(x^3 + x + 1)(x^3 + x^2 + 1)$$

and $\qquad x^{31} - 1 = (x - 1)g(x)\bar{g}(x),$

where $g(x) = 1 + x^3 + x^8 + x^9 + x^{13} + x^{14} + x^{15}$.

In view of Remark 12.16, we prove the next two lemmas for p equal to a general odd prime number rather than just for $p = 23$. We will denote the vector $1 + x + x^2 + \cdots + x^{p-1}$ consisting of all 1s by **1**. Note that if $x^p - 1 = (x - 1)g_1(x)g_2(x)$, then $g_1(x)g_2(x) = \mathbf{1}$.

Lemma 12.17 Suppose that $x^p - 1 = (x - 1)g_1(x)g_2(x)$ over $GF(2)$, and that $\langle g_1(x) \rangle$ and $\langle g_2(x) \rangle$ are equivalent codes. Let $a(x)$ be a codeword of $\langle g_1(x) \rangle$ of *odd* weight w. Then
(i) $w^2 \geq p$
(ii) if also $g_2(x) = \bar{g}_1(x)$, then $w^2 - w + 1 \geq p$.

Proof (i) Since $\langle g_2(x) \rangle$ is equivalent to $\langle g_1(x) \rangle$, there is some codeword $b(x)$ in $\langle g_2(x) \rangle$ also of weight w. Now $a(x)b(x)$ is a multiple of $g_1(x)g_2(x) = \mathbf{1}$, and so $a(x)b(x) = 0$ or **1**. Since w is odd, we have $a(1)b(1) = w \cdot w \equiv 1 \bmod (2)$, and so we must have $a(x)b(x) = 1 + x + \cdots + x^{p-1}$. But $a(x)b(x)$ has at most w^2 non-zero coefficients and so $w^2 \geq p$.
  (ii) If $g_2(x) = \bar{g}_1(x)$, then the codewords of $\langle g_2(x) \rangle$ are just the reciprocals of the codewords of $\langle g_1(x) \rangle$. In particular we may take $b(x)$ to be $a(x^{-1})$ in the proof of (i) to get

$$a(x)a(x^{-1}) = 1 + x + x^2 + \cdots + x^{p-1}.$$

But w of the w^2 terms in the product $a(x)a(x^{-1})$ are 1 and so the maximum weight of $a(x)a(x^{-1})$ is $w^2 - w + 1$.

Corollary 12.18 If, with the hypotheses of Lemma 12.17, it is also known that the minimum distance d of $\langle g_1(x)\rangle$ is *odd*, then d satisfies the *square root bound*

$$d \geqslant \sqrt{p},$$

while if also $g_2(x) = \bar{g}_1(x)$, this can be improved to

$$d^2 - d + 1 \geqslant p.$$

By Lemma 12.17(ii), our $[23, 12]$-code C_1 has no words of *odd* weight less than 7, because $5^2 - 5 + 1 < 23$. There is an ingenious way of showing that C_1, and more generally any so-called quadratic residue (QR) code (we do not define QR codes here, but simply remark that C_1 is an example of such a code), must have odd minimum distance and therefore must satisfy the square root bound. The argument, which involves showing that an extended QR code has a transitive automorphism group, is beyond the scope of the present book. As our main aim is merely to find the minimum distance of the Golay code C_1, the following lemma will suffice.

Lemma 12.19 Suppose p is an odd prime number and that, over $GF(2)$, $x^p - 1 = (x-1)g_1(x)\bar{g}_1(x)$. Let $a(x)$ be a codeword of $\langle g_1(x)\rangle$ of even weight w. Then
(i) $w \equiv 0 \pmod 4$
(ii) $w \neq 4$ unless $p = 7$.

Proof (i) As in the proof of Lemma 12.17, we have $a(x)a(x^{-1}) = 0$ or 1. Since $a(x)$ has even weight, $a(1) = 0$, and so $a(x)a(x^{-1}) = 0$. Suppose $a(x) = x^{e_1} + x^{e_2} + \cdots + x^{e_w}$. Then

$$a(x)a(x^{-1}) = \sum_{i=1}^{w} \sum_{j=1}^{w} x^{e_i - e_j} = 0$$

in R_p. Of the w^2 summands, w are equal to 1 (the terms with $i = j$), and these sum to $0 \pmod 2$. So the remaining $w^2 - w$ terms $x^{e_i - e_j} (i \neq j)$ must cancel each other out in pairs. Now if $x^{e_i - e_j} = x^{e_k - e_l}$ then $x^{e_j - e_i} = x^{e_l - e_k}$, and so the terms must actually cancel four at a time. Thus

$$w^2 - w \equiv 0 \pmod 4 \qquad \text{and so} \qquad w \equiv 0 \pmod 4.$$

(ii) Suppose $w = 4$. Without loss of generality (via a suitable cyclic shift), suppose $a(x) = 1 + x^i + x^j + x^k$, where i, j, k are distinct and $1 < i, j, k < p$. Then $(1 + x^i + x^j + x^k)(1 + x^{-i} + x^{-j} + x^{-k}) = 0$.

Thus the six sets $\{i, -i\}$, $\{j, -j\}$, $\{k, -k\}$, $\{i - j, j - i\}$, $\{i - k, k - i\}$ and $\{j - k, k - j\}$ must split into three matching pairs, under congruence modulo p. By symmetry there is no loss in assuming i is congruent to one of $-j$, $j - i$ or $j - k$.

Case 1 Suppose $i \equiv j - k \pmod{p}$. Then $k \equiv j - i$ gives a second match and so the third match must be given by $j \equiv \pm(i - k)$. But $i \equiv j - k$ and $j \equiv i - k$ implies $2k \equiv 0 \pmod{p}$, which is a contradiction since p is an odd prime. Likewise, $i \equiv j - k$ and $j \equiv k - i$ implies $2i \equiv 0 \pmod{p}$, which is again a contradiction.

Case 2 Suppose $i \equiv -j \pmod{p}$. Since Case 1 has been ruled out, we must have $k \equiv i - k$ or $k \equiv j - k$ and as the two possibilities are essentially the same, we may assume $k \equiv i - k$, i.e. $i \equiv 2k$. The third match is then given by $i - j \equiv j - k$, which implies $k \equiv -3i \equiv -6k$. Thus $7k \equiv 0 \pmod{p}$, which is a contradiction unless $p = 7$.

Case 3 Suppose $i \equiv j - i \pmod{p}$. To avoid the cases above, we may assume the remaining matches are given by $j \equiv k - j$ and $k \equiv i - k$. But then $k \equiv 2j \equiv 4i \equiv 8k$, again giving $7k \equiv 0 \pmod{p}$.

Remark We observed in Remark 12.16 that $x^7 - 1$ has the form $(x - 1)g(x)\bar{g}(x)$, where $g(x) = x^3 + x + 1$. Since $\langle g(x) \rangle$ contains words of weight 4, the exclusion of case $p = 7$ in Lemma 12.19(ii) is essential.

We have now reached our goal:

Theorem 12.20 Let G_{23} be the binary cyclic code in R_{23} with generator polynomial $g(x) = 1 + x^2 + x^4 + x^5 + x^6 + x^{10} + x^{11}$. Then G_{23} is a perfect [23, 12, 7]-code.

Proof We have already observed that

$$x^{23} - 1 = (x - 1)g(x)\bar{g}(x).$$

By Lemma 12.17, the minimum *odd* weight w of G_{23} satisfies $w^2 - w + 1 \geq 23$, which implies that $w \geq 7$. By Lemma 12.19, G_{23} can have no words of *even* weight < 8. As $g(x)$ is a codeword of weight 7, we have $d(G_{23}) = 7$. Since

$$2^{12}\left\{1 + 23 + \binom{23}{2} + \binom{23}{3}\right\} = 2^{23},$$

the sphere-packing condition (9.1) is satisfied and so G_{23} is perfect.

The code G_{23} is called the *binary Golay code*. It is equivalent to the Golay code as defined in Chapter 9 (cf. the remarks following Problem 9.9).

The ternary Golay code

We now show that the ternary Golay code G_{11} may also be constructed as a cyclic code. Our starting point is the factorization of $x^{11} - 1$ over $GF(3)$:

$$x^{11} - 1 = (x - 1)(x^5 + x^4 - x^3 + x^2 - 1)(x^5 - x^3 + x^2 - x - 1)$$
$$= (x - 1)g_1(x)g_2(x), \text{ say.}$$

Note that $g_2(x) = -x^5 g_1(x^{-1})$ and so $\langle g_1(x)\rangle$ and $\langle g_2(x)\rangle$ are equivalent [11, 6]-codes. We shall show that the code $\langle g_1(x)\rangle$ has minimum distance 5.

Theorem 12.21 Let C be the ternary code $\langle g_1(x)\rangle$ in R_{11}, where $g_1(x) = x^5 + x^4 - x^3 + x^2 - 1$. Let D be the subcode of C generated by $(x - 1)g_1(x)$. Let $a(x) = a_0 + a_1 x + \cdots + a_{10}x^{10}$ be a codeword of C of weight w. Then
  (i) $a(x) \in D$ if and only if $\sum_{i=0}^{10} a_i = 0$,
  (ii) if $a(x) \in D$, then $w \equiv 0 \pmod{3}$,
  (iii) if $a(x) \notin D$, then $w \equiv 2 \pmod 3$,
  (iv) if $a(x) \notin D$, then $w \geq 4$,
  (v) $w \neq 3$,
  (vi) $d(C) = 5$.

Proof (i)   Given that $a(x)$ is in C and so is a multiple of $g_1(x)$, we have

$$a(x) \in D \Leftrightarrow a(x) \text{ is a multiple of } (x - 1)$$
$$\Leftrightarrow a(1) = 0$$
$$\Leftrightarrow \sum_{i=0}^{10} a_i = 0.$$

(ii)   First observe that, since $a_i^2 \equiv 1 \pmod 3$ for each non-zero coefficient a_i, we have $w \equiv \Sigma\, a_i^2 \pmod 3$. By Theorem 12.15(ii), the dual code $D^\perp$ of D is generated by the reciprocal polynomial of $g_2(x)$, which happens to be precisely $-g_1(x)$. Thus $D^\perp = \langle \bar{g}_2(x) \rangle = \langle -g_1(x) \rangle = C$. So D is contained in $D^\perp$, which means that D is self-orthogonal, i.e. the inner product of any two vectors of D is zero. In particular, if $a(x) \in D$, then the inner product of $a(x)$ with itself is zero, i.e. $\Sigma\, a_i^2 \equiv 0 \pmod 3$. Thus $a(x) \in D \Rightarrow w \equiv 0 \pmod 3$.

(iii)   By Theorem 12.12, D is a code of dimension 5. Also D is contained within the 6-dimensional code C. Since $\mathbf{1} = 1 + x + \cdots + x^{10}$ is in C but not in D, C is the disjoint union of the three cosets $D, \mathbf{1} + D$ and $-\mathbf{1} + D$. Thus any codeword $a(x)$ of C which is not in D is of the form

$$a(x) = d(x) \pm \mathbf{1},$$

for some codeword $d(x) = d_0 + d_1 x + \cdots + d_{10} x^{10} \in D$.

Hence $w(a(x)) \equiv \displaystyle\sum_{i=0}^{10} (d_i \pm 1)^2$

$$\equiv \left(\sum_{i=0}^{10} d_i^2\right) + 11 \pm 2\left(\sum_{i=0}^{10} d_i\right)$$

$$\equiv 11 \qquad \text{(by (i) and (ii))}$$

$$\equiv 2 \pmod 3.$$

(iv)   Suppose $a(x) \notin D$. Now $a(x)a(x^{-1})$ is a multiple of $g_1(x)g_2(x) = \mathbf{1}$. By (i), $a(1) \neq 0$, and so $a(x)a(x^{-1}) = \pm \mathbf{1}$. Thus $a(x)a(x^{-1})$ has weight 11. But at most w^2 coefficients of $a(x)a(x^{-1})$ are non-zero and so $w^2 \geq 11$. Hence $w \geq 4$.

(v)   Suppose, for a contradiction, that $w = 3$. Then, by a suitable cyclic shift, and multiplication by -1 if necessary, we may suppose $a(x) = 1 \pm x^i \pm x^j$. By (ii) and (iii), $a(x)$ must be in D and so, by (i), we must actually have $a(x) = 1 + x^i + x^j$. Also, $a(x) \in D$ implies that $a(x)a(x^{-1})$ is a multiple of

$$(x - 1)g_1(x)g_2(x) = x^{11} - 1 = 0$$

in R_{11}. Thus

$$(1 + x^i + x^j)(1 + x^{-i} + x^{-j}) = 0,$$

giving $x^i + x^{-i} + x^j + x^{-j} + x^{j-i} + x^{i-j} = 0.$

Since i and j are distinct and non-zero we must have $i \equiv -j \equiv j - i \pmod{11}$, which implies that $3j \equiv 0 \pmod{11}$, which is a contradiction.

(vi) It follows from (ii)–(v) that $d(C) \geq 5$ and since $g_1(x)$ itself has weight 5, $d(C) = 5$.

The $[11, 6, 5]$ code C of Theorem 12.21 is called the *ternary Golay code*. It is a perfect code because

$$3^6\left\{1 + 2 \cdot 11 + 2^2\binom{11}{2}\right\} = 3^{11},$$

and it is equivalent to the ternary Golay code defined in Chapter 9.

Hamming codes as cyclic codes

We will show that the binary Hamming codes discussed in Chapter 8 are equivalent to cyclic codes. The proof will be incomplete in the sense that we shall assume results previously stated, but left unproved, in the text.

Theorem 12.22 The binary Hamming code Ham $(r, 2)$ is equivalent to a cyclic code.

Proof Let $p(x)$ be an irreducible polynomial of degree r in $F_2[x]$. Then, by Theorem 12.5, the ring $F_2[x]/p(x)$ of polynomials modulo $p(x)$ is actually a field of order 2^r. As was mentioned in Chapter 3, every finite field has a primitive element and so there exists an element α of $F_2[x]/p(x)$ such that $F_2[x]/p(x) = \{0, 1, \alpha, \alpha^2, \ldots, \alpha^{2^r-2}\}$. Let us now identify an element $a_0 + a_1x + a_2x^2 + \cdots + a_{r-1}x^{r-1}$ of $F_2[x]/p(x)$ with the column vector

$$\begin{bmatrix} a_0 \\ a_1 \\ \vdots \\ a_{r-1} \end{bmatrix}$$

and consider the binary $r \times (2^r - 1)$ matrix

$$H = [1 \ \alpha \ \alpha^2 \cdots \alpha^{2^r-2}].$$

Let C be the binary linear code having H as parity-check matrix.

Since the columns of H are precisely the distinct non-zero vectors of $V(r, 2)$, C is a Hamming code Ham $(r, 2)$. Putting $n = 2^r - 1$ we have

$$C = \{f_0 f_1 \cdots f_{n-1} \in V(n, 2) \mid f_0 + f_1 \alpha + \cdots + f_{n-1}\alpha^{n-1} = 0\}$$
$$= \{f(x) \in R_n \mid f(\alpha) = 0 \text{ in } F_2[x]/p(x)\}. \tag{12.23}$$

If $f(x) \in C$ and $r(x) \in R_n$, then $r(x)f(x) \in C$ because $r(\alpha)f(\alpha) = r(\alpha) \cdot 0 = 0$. So, by Theorem 12.6, this version of Ham $(r, 2)$ is cyclic.

Definition If $p(x)$ is an irreducible polynomial of degree r such that x is a primitive element of the field $F[x]/p(x)$, then $p(x)$ is called a *primitive polynomial*.

Theorem 12.24 If $p(x)$ is a primitive polynomial over $GF(2)$ of degree r, then the cyclic code $\langle p(x) \rangle$ is the Hamming code Ham $(r, 2)$.

Proof If $p(x)$ is primitive, then (12.23) implies that

$$\text{Ham } (r, 2) = \{f(x) \in R_n \mid f(x) = 0 \text{ in } F_2[x]/p(x)\}$$
$$= \langle p(x) \rangle.$$

Example 12.25 The polynomial $x^3 + x + 1$ is irreducible over $GF(2)$ and so $F_2[x]/(x^3 + x + 1)$ is a field of order 8. Also, x is a primitive element of this field, for

$$F_2[x]/(x^3 + x + 1)$$
$$= \{0, 1, x, x^2, x^3 = x + 1, x^4 = x^2 + x, x^5 = x^2 + x + 1, x^6 = x^2 + 1\}.$$

Thus a parity-check matrix for a cyclic version of the Hamming code Ham $(3, 2)$ is

$$H = \begin{bmatrix} 1 & 0 & 0 & 1 & 0 & 1 & 1 \\ 0 & 1 & 0 & 1 & 1 & 1 & 0 \\ 0 & 0 & 1 & 0 & 1 & 1 & 1 \end{bmatrix},$$

wherein the columns represent $1, \alpha, \alpha^2, \ldots, \alpha^6$ as described in the proof of Theorem 12.22, with $\alpha = x$.

Since $x^3 + x + 1$ is a primitive polynomial, it is a generator polynomial for Ham $(3, 2)$ and so, by Theorem 12.12, a generator matrix for the code is

$$G = \begin{bmatrix} 1 & 1 & 0 & 1 & 0 & 0 & 0 \\ 0 & 1 & 1 & 0 & 1 & 0 & 0 \\ 0 & 0 & 1 & 1 & 0 & 1 & 0 \\ 0 & 0 & 0 & 1 & 1 & 0 & 1 \end{bmatrix}.$$

Remarks (1) It can be shown that there exists a primitive polynomial of degree r for any r.

(2) We saw in Example 12.13 that the ternary Hamming code Ham $(2, 3)$ is not equivalent to a cyclic code. However, Ham (r, q) is equivalent to a cyclic code if r and $q - 1$ are relatively prime (see, e.g., Blahut (1983), Theorem 5.5.1).

Concluding remarks on Chapter 12

(1) Cyclic codes were first studied by Prange (1957). Interest was further stimulated by the theorem of Bose and Ray-Chaudhuri (1960) which gave lower bounds on the minimum distance for a large class of cyclic codes. It was quickly discovered that almost every special linear code previously discovered (e.g. Hamming, Golay, Reed–Muller) could be made cyclic.

(2) For a comprehensive treatment of the theory of cyclic codes, see, e.g., MacWilliams and Sloane (1977). For details of the practical implementation of cyclic codes, including the associated circuitry, see, e.g., Blahut (1983) or Lin and Costello (1983).

Exercises 12

12.1 Is each of the following codes (a) cyclic, (b) equivalent to a cyclic code?
   (i)   the binary code {0000, 1100, 0110, 0011, 1001}
   (ii) the binary code {00000, 10110, 01101, 11011}
   (iii) the ternary code {0000, 1122, 2211}
   (iv) the q-ary repetition code of length n
   (v)   the binary even-weight code E_n
   (vi) the ternary code $\{\mathbf{x} \in V(n, 3) \mid w(\mathbf{x}) \equiv 0 \,(\text{mod } 3)\}$

(vii)   the ternary code

$$\left\{ x_1 x_2 \cdots x_n \in V(n, 3) \,\middle|\, \sum_{i=1}^{n} x_i \equiv 0 \pmod 3 \right\}$$

12.2   Write out the multiplication table for $F_2[x]/(x^2 + 1)$. Explain why $F_2[x]/(x^2 + 1)$ is not a field.

12.3   Write out a proof of Theorem 12.5.

12.4   Show that an irreducible polynomial over $GF(2)$ of degree ≥ 2 has an odd number of non-zero coefficients.

12.5   To verify that a polynomial $p(x)$ is irreducible, why is it enough to show that $p(x)$ has no irreducible factor of degree $\leq \frac{1}{2} \deg p(x)$?

12.6   List the irreducible polynomials over $GF(2)$ of degrees 1 to 4. Construct a finite field of order 8.

12.7   Suppose p is a prime number.
   (i)   Factorize $x^p - 1$ into irreducible polynomials over $GF(p)$.
   (ii) Factorize $x^{p-1} - 1$ into irreducible polynomials over $GF(p)$.

12.8   Factorize $x^5 - 1$ into irreducible polynomials over $GF(2)$ and hence determine all the cyclic binary codes of length 5.

12.9   Let $g(x)$ be the generator polynomial of a binary cyclic code which contains some codewords of odd weight. Is the set of codewords in $\langle g(x) \rangle$ of even weight a cyclic code? If so, what is the generator polynomial of this subcode?

12.10 Suppose $x^n - 1$ is the product of t distinct irreducible polynomials over $GF(q)$. How many cyclic codes of length n over $GF(q)$ are there?

12.11 Given that the factorization of $x^7 - 1$ into irreducible polynomials over $GF(2)$ is $(x - 1)(x^3 + x + 1)(x^3 + x^2 + 1)$, determine all the cyclic binary codes of length 7. Give a name or a concise description of each of these codes.

12.12 Factorize $x^8 - 1$ over $GF(3)$. How many ternary cyclic codes of length 8 are there?

12.13 Write down a check polynomial and a parity-check matrix for each of the ternary cyclic codes of length 4 (see Example 12.13).

12.14 Let $h(x)$ be the check polynomial of a cyclic code C. Is $\langle h(x) \rangle$ equal to $C^\perp$? Is $\langle h(x) \rangle$ equivalent to $C^\perp$?

12.15 Suppose C is a binary cyclic code of odd length. Show that C contains a codeword of odd weight if and only if **1** is a codeword of C.

12.16 Suppose a generator matrix G of a linear code C has the property that a cyclic shift of any row of G is also a codeword. Show that C is a cyclic code.

12.17 Show that 2 is a primitive element of $GF(11)$. Deduce that the $[10, 8]$- and $[10, 6]$-codes over $GF(11)$ of Examples 7.12 and 11.3 respectively are equivalent to cyclic codes.

12.18 Let G_{23} be the cyclic Golay code defined in the text. Prove that any two vectors in G_{23} of even weight have inner product equal to zero. Hence prove that the extended Golay code G_{24}, obtained by adding an overall parity-check to G_{23}, is self-dual.

12.19 Determine which of the irreducible polynomials over $GF(2)$ of degree 4 (found in Exercise 12.6) are primitive. Hence write down a generator polynomial for the binary Hamming code of length 15. Find the check polynomial for this code. Write down the corresponding parity-check matrix (using Theorem 12.15) and check that its columns are precisely the non-zero vectors of $V(4, 2)$.

12.20 Let $g(x)$ be the generator polynomial of a cyclic binary Hamming code Ham $(r, 2)$, with $r \geqslant 3$. Show that $\langle (x - 1)g(x) \rangle$ is a cyclic $[2^r - 1, 2^r - r - 2, 4]$-code.

12.21 An error vector of the form $x^i + x^{i+1}$ in R_n is called a *double-adjacent error*. Show that the code $\langle (x - 1)g(x) \rangle$ of Exercise 12.20 is capable of correcting all single errors and all double-adjacent errors.

12.22 Let C be a $[q + 1, 2, q]$-code over $GF(q)$, where q is odd. Show that C cannot be cyclic. Deduce that the Hamming code Ham $(2, q)$ is not equivalent to a cyclic code when q is odd.

13 Weight enumerators

If C is a linear $[n, k]$-code, its *weight enumerator* is defined to be the polynomial

$$W_C(z) = \sum_{i=0}^{n} A_i z^i$$

$$= A_0 + A_1 z + \cdots + A_n z^n,$$

where A_i denotes the number of codewords in C of weight i.
    Another way of writing $W_C(z)$ is

$$W_C(z) = \sum_{\mathbf{x} \in C} z^{w(\mathbf{x})}.$$

Examples 13.1 (i)   Let C be the binary even-weight code of length 3; i.e. $C = \{000, 011, 101, 110\}$. Its dual code $C^{\perp}$ is $\{000, 111\}$. The weight enumerators of C and $C^{\perp}$ are

$$W_C(z) = 1 + 3z^2$$
$$W_{C^{\perp}}(z) = 1 + z^3.$$

   (ii)   The code $C = \{00, 11\}$ is self-dual and so

$$W_C(z) = W_{C^{\perp}}(z) = 1 + z^2.$$

We have already seen (Theorem 6.14) that knowledge of the weight enumerator of a code enables us to calculate the probability of undetected errors when the code is used purely for error detection.
    The main result of this chapter is a remarkable formula of MacWilliams (1963), which enables the weight enumerator of any linear code C to be obtained from the weight enumerator of its dual code $C^{\perp}$.
    For simplicity we shall prove this result, known as the MacWilliams identity, only for binary codes (Theorem 13.5), although the general result will be stated afterwards (Theorem 13.6).
    The following three lemmas are required only for the proof of

the MacWilliams identity. The less mathematically minded reader, who is happy to accept the validity of the formula without proof, may skip these lemmas, and also the proof of Theorem 13.5, without any great loss; the subsequent examples and exercises make use only of the formula and not of its proof.

Lemma 13.2 Let C be a binary linear $[n, k]$-code and suppose $\mathbf{y}$ is a fixed vector in $V(n, 2)$ which is not in $C^\perp$. Then $\mathbf{x} \cdot \mathbf{y}$ is equal to 0 and 1 equally often as $\mathbf{x}$ runs over the codewords of C.

Proof Let $A = \{\mathbf{x} \in C \mid \mathbf{x} \cdot \mathbf{y} = 0\}$
and $B = \{\mathbf{x} \in C \mid \mathbf{x} \cdot \mathbf{y} = 1\}$.

Let $\mathbf{u}$ be a codeword of C such that $\mathbf{u} \cdot \mathbf{y} = 1$ ($\mathbf{u}$ exists since $\mathbf{y} \notin C^\perp$). Let $\mathbf{u} + A$ denote the set $\{\mathbf{u} + \mathbf{x} \mid \mathbf{x} \in A\}$. Then

$$\mathbf{u} + A \subseteq B,$$

for if $\mathbf{x} \in A$, then $(\mathbf{u} + \mathbf{x}) \cdot \mathbf{y} = \mathbf{u} \cdot \mathbf{y} + \mathbf{x} \cdot \mathbf{y} = 1 + 0 = 1$.
  Similarly

$$\mathbf{u} + B \subseteq A.$$

Hence

$$|A| = |\mathbf{u} + A| \le |B| = |\mathbf{u} + B| \le |A|.$$

Hence $|A| = |B|$ and the lemma is proved.

Lemma 13.3 Let C be a binary $[n, k]$-code and let $\mathbf{y}$ be any element of $V(n, 2)$. Then

$$\sum_{\mathbf{x} \in C} (-1)^{\mathbf{x} \cdot \mathbf{y}} = \begin{cases} 2^k & \text{if } \mathbf{y} \in C^\perp \\ 0 & \text{if } \mathbf{y} \notin C^\perp. \end{cases}$$

Proof If $\mathbf{y} \in C^\perp$, then $\mathbf{x} \cdot \mathbf{y} = 0$ for all $\mathbf{x} \in C$, and so

$$\sum_{\mathbf{x} \in C} (-1)^{\mathbf{x} \cdot \mathbf{y}} = |C| \cdot 1 = 2^k.$$

If $\mathbf{y} \notin C^\perp$, then by Lemma 13.2, as $\mathbf{x}$ runs over the elements of C, $(-1)^{\mathbf{x} \cdot \mathbf{y}}$ is equal to 1 and -1 equally often, giving

$$\sum_{\mathbf{x} \in C} (-1)^{\mathbf{x} \cdot \mathbf{y}} = 0.$$

Lemma 13.4 Let **x** be a fixed vector in $V(n, 2)$ and let z be an indeterminate. Then the following polynomial identity holds:

$$\sum_{\mathbf{y} \in V(n,2)} z^{w(\mathbf{y})}(-1)^{\mathbf{x} \cdot \mathbf{y}} = (1 - z)^{w(\mathbf{x})}(1 + z)^{n - w(\mathbf{x})}.$$

Proof

$$\sum_{\mathbf{y} \in V(n,2)} z^{w(\mathbf{y})}(-1)^{\mathbf{x} \cdot \mathbf{y}} = \sum_{y_1=0}^{1} \sum_{y_2=0}^{1} \cdots \sum_{y_n=0}^{1} z^{y_1+y_2+\cdots+y_n}(-1)^{x_1 y_1 + \cdots + x_n y_n}$$

$$= \sum_{y_1=0}^{1} \cdots \sum_{y_n=0}^{1} \left(\prod_{i=1}^{n} z^{y_i}(-1)^{x_i y_i} \right)$$

$$= \prod_{i=1}^{n} \left(\sum_{j=0}^{1} z^{j}(-1)^{j x_i} \right)$$

$$= (1 - z)^{w(\mathbf{x})}(1 + z)^{n - w(\mathbf{x})},$$

since
$$\sum_{j=0}^{1} z^{j}(-1)^{j x_i} = \begin{cases} 1 + z & \text{if } x_i = 0 \\ 1 - z & \text{if } x_i = 1 \end{cases}.$$

Theorem 13.5 (*The MacWilliams identity for binary linear codes*) If C is a binary $[n, k]$-code with dual code $C^{\perp}$, then

$$W_{C^{\perp}}(z) = \frac{1}{2^k}(1 + z)^n W_C \left(\frac{1 - z}{1 + z} \right).$$

Proof We express the polynomial

$$f(z) = \sum_{\mathbf{x} \in C} \left(\sum_{\mathbf{y} \in V(n,2)}' (-1)^{\mathbf{x} \cdot \mathbf{y}} z^{w(\mathbf{y})} \right)$$

in two ways.

On the one hand, using Lemma 13.4,

$$f(z) = \sum_{\mathbf{x} \in C} (1 - z)^{w(\mathbf{x})}(1 + z)^{n - w(\mathbf{x})}$$

$$= (1 + z)^n \sum_{\mathbf{x} \in C} \left(\frac{1 - z}{1 + z} \right)^{w(\mathbf{x})}$$

$$= (1 + z)^n W_C \left(\frac{1 - z}{1 + z} \right).$$

On the other hand, reversing the order of summation, we have

$$f(z) = \sum_{y \in V(n,2)} z^{w(y)} \left(\sum_{x \in C} (-1)^{x \cdot y} \right)$$

$$= \sum_{y \in C^{\perp}} z^{w(y)} 2^k \quad \text{(by Lemma 13.3)}$$

$$= 2^k W_{C^{\perp}}(z).$$

Equating the two expressions for $f(z)$ establishes the result.

The proof of the following more general result is similar to that of Theorem 13.5, using generalized versions of the preceding lemmas, but we omit the details.

Theorem 13.6 (*The MacWilliams identity for general linear codes*) If C is a linear $[n, k]$-code over $GF(q)$ with dual code $C^{\perp}$, then

$$W_{C^{\perp}}(z) = \frac{1}{q^k} [1 + (q-1)z]^n W_C \left(\frac{1-z}{1+(q-1)z} \right).$$

Remark If C is a binary $[n, k]$-code, then, since the dual code of $C^{\perp}$ is just C, we can write the MacWilliams identity in the (often more useful) form:

$$W_C(z) = \frac{1}{2^{n-k}} (1+z)^n W_{C^{\perp}} \left(\frac{1-z}{1+z} \right). \tag{13.7}$$

Examples 13.8 We apply Theorem 13.5 to the codes of Examples 13.1.
  (i) We have $W_C(z) = 1 + 3z^2$. Hence, by Theorem 13.5,

$$W_{C^{\perp}}(z) = \tfrac{1}{4}(1+z)^3 W_C \left(\frac{1-z}{1+z} \right)$$

$$= \tfrac{1}{4}[(1+z)^3 + 3(1-z)^2(1+z)]$$

$$= 1 + z^3,$$

as already found directly from $C^{\perp}$.
  Let us interchange the roles of C and $C^{\perp}$ in order to check the

formula (13.7). We have

$$\tfrac{1}{2}(1+z)^3 W_{C^\perp}\left(\frac{1-z}{1+z}\right) = \tfrac{1}{2}[(1+z)^3 + (1-z)^3]$$

$$= 1 + 3z^2,$$

which is indeed $W_C(z)$.

(ii) We have $W_C(z) = 1 + z^2$. Hence

$$W_{C^\perp}(z) = \tfrac{1}{2}(1+z)^2 W_C\left(\frac{1-z}{1+z}\right)$$

$$= \tfrac{1}{2}[(1+z)^2 + (1-z)^2]$$

$$= 1 + z^2.$$

Thus $W_{C^\perp}(z) = W_C(z)$, as we expect, since C is self-dual.

For the very small codes just considered, the use of the MacWilliams identity is an inefficient way of calculating their weight enumerators, which can be written down directly from the lists of codewords. But suppose we are required to calculate the weight enumerator of an $[n, k]$-code C over $GF(q)$ where k is large. To enumerate all q^k codewords by weight may be a formidable task. However, if k is so large that $n - k$ is small, then the dual code $C^\perp$ may be small enough to find *its* weight enumerator, and then the MacWilliams identity can be used to find the weight enumerator of C.

For example, the binary Hamming code Ham $(r, 2)$ has dimension $2^r - 1 - r$, and so the number of codewords in Ham $(r, 2)$ is 2^{2^r-1-r}, a large number even for moderately small values of r. But the dual code has only 2^r codewords and, as we shall soon see, it has a particularly simple weight enumerator. From this, the weight enumerator of Ham $(r, 2)$ itself is easily determined. First we look at a particular case.

Example 13.9 Let C be the binary $[7, 4]$-Hamming code. Then the dual code $C^\perp$ has generator matrix

$$\begin{bmatrix} 0 & 0 & 0 & 1 & 1 & 1 & 1 \\ 0 & 1 & 1 & 0 & 0 & 1 & 1 \\ 1 & 0 & 1 & 0 & 1 & 0 & 1 \end{bmatrix}.$$

When we compute $W_{C^\perp}(z)$ directly, by listing the codewords, we find, surprisingly, that each of the non-zero codewords has weight 4 (the next theorem shows this to be no isolated phenomenon, as far as the Hamming codes are concerned). Thus

$$W_{C^\perp}(z) = 1 + 7z^4,$$

and so the weight enumerator of C itself is, by equation (13.7),

$$\tfrac{1}{8}[(1+z)^7 + 7(1-z)^4(1+z)^3] = 1 + 7z^3 + 7z^4 + z^7.$$

Theorem 13.10 Let C be the binary Hamming code Ham $(r, 2)$. Then every non-zero codeword of $C^\perp$ has weight 2^{r-1}.

Proof Let

$$H = \begin{bmatrix} \mathbf{h}_1 \\ \mathbf{h}_2 \\ \vdots \\ \mathbf{h}_r \end{bmatrix} = \begin{bmatrix} h_{11} & h_{12} & \cdots & h_{1n} \\ h_{21} & h_{22} & \cdots & h_{2n} \\ \vdots & \vdots & & \vdots \\ h_{r1} & h_{r2} & \cdots & h_{rn} \end{bmatrix}$$

be a parity-check matrix of C where the rows of H are denoted by $\mathbf{h}_1, \mathbf{h}_2, \ldots, \mathbf{h}_r$. Then a non-zero codeword $\mathbf{c}$ of $C^\perp$ is a vector of the form $\mathbf{c} = \sum_{i=1}^{r} \lambda_i \mathbf{h}_i$ for some scalars $\lambda_1, \lambda_2, \ldots, \lambda_r$, not all zero. We will find the weight of $\mathbf{c}$ by finding the number $n_0(\mathbf{c})$ of *zero* entries of $\mathbf{c}$ and then subtracting $n_0(\mathbf{c})$ from the length n. Now $\mathbf{c}$ has a zero in its jth position if and only if $\sum_{i=1}^{r} \lambda_i h_{ij} = 0$, i.e. if and only if $\sum_{i=1}^{r} \lambda_i x_i = 0$, where $(x_1 x_2 \cdots x_r)^T$ is the jth column of H. Since C is a Hamming code, the columns of H are precisely the non-zero vectors of $V(r, 2)$ and so $n_0(\mathbf{c})$ is equal to the number of *non-zero* vectors in the set

$$X = \left\{ x_1 x_2 \cdots x_r \in V(r, 2) \,\middle|\, \sum_{i=1}^{r} \lambda_i x_i = 0 \right\},$$

i.e. $n_0(\mathbf{c}) = |X| - 1$.

It is easy to see that X is an $(r-1)$-dimensional subspace of $V(r, 2)$ (e.g. view X as the dual code of the $[r, 1]$-code which has generator matrix $[\lambda_1 \lambda_2 \cdots \lambda_r]$, so that $\dim (X) = r - 1$, by Theorem 7.3). Hence

$$|X| = 2^{r-1} \quad \text{and so} \quad n_0(\mathbf{c}) = 2^{r-1} - 1.$$

(Note that $n_0(\mathbf{c})$ is independent of the choice of non-zero codeword $\mathbf{c}$ in $C^\perp$). Thus

$$w(\mathbf{c}) = n - n_0(\mathbf{c}) = 2^r - 1 - (2^{r-1} - 1)$$
$$= 2^{r-1}.$$

Corollary 13.11 The weight enumerator of the binary Hamming code Ham $(r, 2)$, of length $n = 2^r - 1$, is

$$\frac{1}{2^r}[(1 + z)^n + n(1 - z^2)^{(n-1)/2}(1 - z)].$$

Proof This is a straightforward application of the MacWilliams identity which is left to Exercise 13.5.

Probability of undetected errors

Suppose we wish to find $P_{\text{undetec}}(C)$ for a binary $[n, k]$-code C. By Theorem 6.14, we have

$$P_{\text{undetec}}(C) = \sum_{i=1}^{n} A_i p^i (1 - p)^{n-i}$$

$$= (1 - p)^n \sum_{i=1}^{n} A_i \left(\frac{p}{1 - p}\right)^i.$$

Since

$$W_C\left(\frac{p}{1 - p}\right) = \sum_{i=0}^{n} A_i \left(\frac{p}{1 - p}\right)^i,$$

and since $A_0 = 1$, we have

$$P_{\text{undetec}}(C) = (1 - p)^n \left[W_C\left(\frac{p}{1 - p}\right) - 1\right]. \qquad (13.12)$$

If we know $W_C(z)$, then we can find $P_{\text{undetec}}(C)$ by means of equation (13.12). If we know only $W_{C^\perp}(z)$ to start with, then we could use the MacWilliams identity (13.7) to calculate $W_C(z)$ and then use equation (13.12). Alternatively, we could use the formula derived in Exercise 13.9, which gives $P_{\text{undetec}}(C)$ directly in terms of $W_{C^\perp}(z)$, and thereby avoid the intermediate calculation of $W_C(z)$.

Exercises 13

13.1 Suppose C is a binary linear code of length n which contains the vector $11 \cdots 1$ consisting of all 1s. Show that

$$A_i = A_{n-i},$$

for $i = 0, 1, \ldots, n$.

13.2 Find the weight enumerator of the code whose generator matrix is

$$\begin{bmatrix} 1 & 0 & 0 & 1 & 1 \\ 0 & 1 & 0 & 0 & 1 \\ 0 & 0 & 1 & 0 & 1 \end{bmatrix}$$

(a) directly,
(b) by using the MacWilliams identity.

13.3 Let C be the binary $[9, 7]$-code having the generator matrix

$$\left[\begin{array}{c|c} & 01 \\ & 01 \\ & 10 \\ I_7 & 10 \\ & 11 \\ & 11 \\ & 11 \end{array} \right].$$

Let $\sum_{i=0}^{9} A_i z^i$ denote the weight enumerator of C. Use the MacWilliams identity to find the values of A_0, A_1, A_2 and A_3. Show that C contains the vector consisting of all 1s and hence, or otherwise, determine the full weight enumerator of C.

13.4 Using the result of Example 13.9, write down the weight enumerator of the extended binary Hamming code of length 8.

13.5 Prove Corollary 13.11.

13.6 Find the number of codewords of each of the weights 0, 1, 2, 3 and 4 in the binary Hamming code of length 15.

13.7 Let C be a binary linear code and let C_0 denote the subcode of C consisting of all codewords of C of even weight. Show that

$$W_{C_0}(z) = \tfrac{1}{2}[W_C(z) + W_C(-z)].$$

13.8 Let C be a binary linear code and let $\hat{C}$ be the extended code obtained from C by adding an overall parity check. Show that

$$W_{\hat{C}}(z) = \tfrac{1}{2}[(1 + z)W_C(z) + (1 - z)W_C(-z)].$$

13.9 Suppose C is a binary $[n, k]$-code. Prove that

$$P_{\text{undetec}}(C) = \frac{1}{2^{n-k}} W_{C^\perp}(1 - 2p) - (1 - p)^n.$$

13.10 Let G_{24} be the extended binary Golay code defined in Theorem 9.3. Notice that the vector consisting of all 1s belongs to G_{24} (add all the rows of G together). Using properties of G_{24} found during the proof of Theorem 9.3, show that

$$W_{G_{24}}(z) = 1 + 759z^8 + 2576z^{12} + 759z^{16} + z^{24}.$$

13.11 Let G_{23} be the cyclic binary code defined in Theorem 12.20, and let G_{24} be its extended code. Using results from Chapter 12, including Exercise 12.18, determine the weight enumerator of G_{24}.

13.12 Use either Exercise 13.10 or 13.11, together with Exercise 9.4(a), to determine the weight enumerator of the binary Golay code G_{23}.

14 The main linear coding theory problem

In Chapter 2 we discussed the 'main coding theory problem'. This was the problem of finding $A_q(n, d)$, the largest value of M for which there exists a q-ary (n, M, d)-code. In the present chapter we shall consider the same problem restricted to linear codes. If q is a prime power, we denote by $B_q(n, d)$ the largest value of M for which there exists a *linear* (n, M, d)-code over $GF(q)$. (The function $B_q(n, d)$ was briefly introduced in Exercises 5.8 and 5.9). Clearly $B_q(n, d)$ is always a power of q, and $B_q(n, d) \leq A_q(n, d)$. We shall refer to the problem of finding $B_q(n, d)$ as the *main linear coding theory problem*, or MLCT problem for short.

If we regard the values of q and d as fixed, we may state the problem as follows.

MLCT problem (Version 1) For given length n, find the maximum dimension k such that there exists an $[n, k, d]$-code over $GF(q)$. (Then, for this k, $B_q(n, d) = q^k$).

Recall that the *redundancy* r of an $[n, k, d]$-code is just $n - k$ (the number of check symbols in a codeword). An alternative version of the MLCT problem is:

MLCT problem (Version 2) For given redundancy r, find the maximum length n such that there exists an $[n, n - r, d]$-code over $GF(q)$.

Solving Version 1 for all n is equivalent to solving Version 2 for all r, because in either case we then know exactly those values of n and k for which an $[n, k, d]$-code exists. The equivalence of the two versions will be made explicit in Theorem 14.3.

It turns out that Version 2 provides the more natural approach. The key to this approach, which was touched upon in

Concluding Remark 3 of Chapter 8, is given in the next theorem. But first we make some definitions.

Definitions An (n, s)-*set* in $V(r, q)$ is a set of n vectors in $V(r, q)$ with the property that any s of them are linearly independent.

We denote by $\max_s (r, q)$ the largest value of n for which there exists an (n, s)-set in $V(r, q)$. An (n, s)-set in $V(r, q)$ which has $n = \max_s (r, q)$ is called *optimal*. The *packing problem* for $V(r, q)$ is that of determining the values of $\max_s (r, q)$ and the optimal (n, s)-sets.

The packing problem was first considered by Bose (1947) for its statistical interest and later (1961) for its connection with coding theory, which is given by the following theorem.

Theorem 14.1 There exists an $[n, n - r, d]$-code over $GF(q)$ if and only if there exists an $(n, d - 1)$-set in $V(r, q)$.

Proof Suppose C is an $[n, n - r, d]$-code over $GF(q)$ with parity-check matrix H. Then, by Theorem 8.4, the columns of H form an $(n, d - 1)$-set in $V(r, q)$. On the other hand, suppose K is an $(n, d - 1)$-set in $V(r, q)$. If we form an $r \times n$ matrix H with the vectors of K as its columns, then, again by Theorem 8.4, H is the parity-check matrix of an $[n, n - r]$-code whose minimum distance is at least d.

Corollary 14.2 For given values of q, d and r, the largest value of n for which there exists an $[n, n - r, d]$-code over $GF(q)$ is $\max_{d-1} (r, q)$.

So the MLCT problem (Version 2) is the same as the packing problem of finding $\max_{d-1} (r, q)$. We now show that the values of $B_q(n, d)$ are also given by the solutions to this problem.

Theorem 14.3 Suppose $\max_{d-1} (r - 1, q) < n \le \max_{d-1} (r, q)$. Then $B_q(n, d) = q^{n-r}$.

Proof Since $n \le \max_{d-1} (r, q)$, there exists an $[n, n - r, d]$-code over $GF(q)$, and so $B_q(n, d) \ge q^{n-r}$. If $B_q(n, d)$ were strictly greater than q^{n-r}, then there would exist an $[n, n - r + 1, d]$-code, implying that $n \le \max_{d-1} (r - 1, q)$, contrary to hypothesis.

Let us pause to outline our plan of campaign for the remainder of this and the next chapter. We shall consider the MLCT problem for increasing values of the minimum distance d. The cases $d = 1$ and $d = 2$ are easily dealt with in Exercise 14.2. We will therefore consider first the problem for $d = 3$ and will solve it for all values of q and r. We will then consider the case $d = 4$, solving the MLCT problem for $q = 2$ and giving the known results for $q > 2$. For cases of d greater than 4, very little is known in the way of general results, at least not until d reaches its maximum value for given redundancy r, which is $d = r + 1$. We will consider this very interesting case in Chapter 15.

The MLCT problem for $d = 3$ (or Hamming codes revisited)

Theorem 14.4 For given redundancy r, the maximum length n of an $[n, n - r, 3]$-code over $GF(q)$ is $(q^r - 1)/(q - 1)$; i.e. $\max_2(r, q) = (q^r - 1)/(q - 1)$.

Proof By Corollary 14.2, the required value of n is $\max_2 (r, q)$, the largest size of an $(n, 2)$-set in $V(r, q)$. Now a set S of vectors in $V(r, q)$ is an $(n, 2)$-set if and only if no vector in S is a scalar multiple of any other vector in S. As we saw in the construction of q-ary Hamming codes in Chapter 8, the $q^r - 1$ non-zero vectors of $V(r, q)$ are partitioned into $(q^r - 1)/(q - 1)$ classes, each class consisting of $q - 1$ vectors which are scalar multiples of each other. Thus an $(n, 2)$-set of largest size is just a set of $(q^r - 1)/(q - 1)$ vectors, one from each of these classes.

The optimal $[n, n - r, 3]$-codes with $n = (q^r - 1)/(q - 1)$ are just the Hamming codes Ham (r, q) defined in Chapter 8. The solution to MLCT problem (Version 1) follows immediately from Theorems 14.3 and 14.4:

Theorem 14.5 $B_q(n, 3) = q^{n-r}$, where r is the unique integer such that $(q^{r-1} - 1)/(q - 1) < n \leq (q^r - 1)/(q - 1)$.

Remarks (1) It is easy to express $B_q(n, 3)$ as an explicit function of q and n (see Exercise 14.3).

(2) To construct a linear $(n, M, 3)$-code with $M = B_q(n, 3)$, one simply finds the least integer r such that $n \leq (q^r - 1)/(q - 1)$ and writes down, as a parity-check matrix, n column vectors of

$V(r, q)$ such that no column is a scalar multiple of another. Such a parity-check matrix can always be obtained by deleting columns from the parity-check matrix of a Hamming code Ham (r, q). Thus the best linear single-error-correcting codes of given length are either Hamming or shortened Hamming codes.

Before proceeding to the case $d = 4$, we remark that it will be advantageous to view an (n, s)-set not only as a set of vectors in the vector space $V(r, q)$, but also as a set of points in the associated projective geometry $PG(r - 1, q)$, which we now define.

The projective geometry $PG(r - 1, q)$

With the vector space $V(r, q) = \{(a_1, a_2, \ldots, a_r) \mid a_i \in GF(q)\}$, we associate a combinatorial structure $PG(r - 1, q)$ consisting of points and lines defined as follows.

The *points* of $PG(r - 1, q)$ are the one-dimensional subspaces of $V(r, q)$. The *lines* of $PG(r - 1, q)$ are the two-dimensional subspaces of $V(r, q)$. The point P is said to belong to (or lie on) the line L if and only if P is a subspace of L. $PG(r - 1, q)$ is called the *projective geometry of dimension $r - 1$ over $GF(q)$*.

Each point P of $PG(r - 1, q)$, as a subspace of $V(r, q)$ of dimension 1, is generated by a single non-zero vector. So, if $\mathbf{a} = (a_1, a_2, \ldots, a_r) \in P$, then

$$P = \{\lambda \mathbf{a} \mid \lambda \in GF(q)\}.$$

In practice, we identify the point P with any non-zero vector it contains. In other words, we take the points of $PG(r - 1, q)$ to be the non-zero vectors of $V(r, q)$ with the rule that if $\mathbf{a} = (a_1, a_2, \ldots, a_r)$ and $\mathbf{b} = (b_1, b_2, \ldots, b_r)$ are two such vectors, then
$\qquad \mathbf{a} = \mathbf{b}$ in $PG(r - 1, q)$ if and only if $\mathbf{a} = \lambda \mathbf{b}$ in $V(r, q)$,

for some non-zero scalar λ.

We now list some elementary properties of $PG(r - 1, q)$.

Lemma 14.6 In $PG(r - 1, q)$,
  (i)   the number of points is $(q^r - 1)/(q - 1)$,
  (ii) any two points lie on exactly one line,
  (iii) each line contains exactly $q + 1$ points,
  (iv) each point lies on $(q^{r-1} - 1)/(q - 1)$ lines.

Proof (i) Since each of the $q^r - 1$ non-zero vectors in $V(r, q)$ has $q - 1$ non-zero scalar multiples, the number of points of $PG(r - 1, q)$ is $(q^r - 1)/(q - 1)$.

(ii) If **a** and **b** are distinct points of $PG(r - 1, q)$, then the unique line through them consists of the points $\lambda \mathbf{a} + \mu \mathbf{b}$, where λ and μ are scalars not both zero.

(iii) In (ii), there are $q^2 - 1$ choices for the pair (λ, μ), but since we are identifying scalar multiples, the number of distinct points on the line is $(q^2 - 1)/(q - 1) = q + 1$.

(iv) Let t be the number of lines on which a given point P lies. Let X denote the set $\{(Q, L) \mid Q$ is a point $\neq P$, L is a line containing both P and $Q\}$. We count the members of X in two ways. For each of the $(q^r - 1)/(q - 1) - 1$ choices for Q, there is a unique line L containing P and Q. Thus

$$|X| = (q^r - 1)/(q - 1) - 1 = (q^r - q)/(q - 1).$$

On the other hand for each of the t lines through P, there are, by part (iii), q points Q other than P lying on L. Thus

$$|X| = tq.$$

Equating the two expressions for $|X|$ gives $t = (q^{r-1} - 1)/(q - 1)$.

Definition The projective geometry $PG(2, q)$ is called the *projective plane over* $GF(q)$. It follows from Lemma 14.6 that $PG(2, q)$ is a symmetric $(q^2 + q + 1, q + 1, 1)$-design, so that it is a projective plane as defined in Chapter 2.

Examples 14.7 (i) The simplest projective plane is $PG(2, 2)$. This contains 7 points labelled $001, 010, 100, 011, 101, 110, 111$,

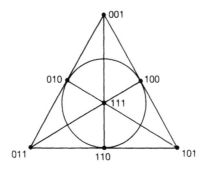

Fig. 14.8. The projective plane $PG(2, 2)$.

and 7 lines as shown in Fig. 14.8. This shows that $PG(2, 2)$ is the same as the 7-point plane of Example 2.19.

(ii)   The 6 points of $PG(1, 5)$ are 01, 10, 11, 12, 13 and 14, and there is just one line consisting of all 6 points. The points could equally well be labelled 03, 10, 22, 12, 21, and 41, say, because in $PG(1, 5)$, $01 = 03$, $11 = 22$, $13 = 21$ and $14 = 41$.

Remarks (1)   The points of $PG(r - 1, q)$ can be uniquely labelled by making the left-most non-zero coordinate equal to 1.

(2)   If $q = 2$, the points of $PG(r - 1, 2)$ are labelled by the non-zero vectors of $V(r, 2)$.

Definition   A set K of n points in $PG(r - 1, q)$ is called an (n, s)-*set* if the vectors representing the points of K form an (n, s)-set in the underlying vector space $V(r, q)$.

Remarks (1)   Two advantages of working in $PG(r - 1, q)$ are that (a) some neat counting arguments may then be used to obtain upper bounds on $\max_s (r, q)$ and (b) many optimal (n, s)-sets turn out to be natural geometric configurations.

(2)   An $(n, 2)$-set in $PG(r - 1, q)$ is just a set of n distinct points of $PG(r - 1, q)$. So we may describe a Hamming code Ham (r, q) as a code having a parity-check matrix H whose columns are the distinct points of $PG(r - 1, q)$. Of course, different representations of these points as vectors will give rise to different, but equivalent, codes. For example (cf. Example 14.7(ii)), Ham $(1, 5)$ may be defined to have parity-check matrix

$$H = \begin{bmatrix} 0 & 1 & 1 & 1 & 1 & 1 \\ 1 & 0 & 1 & 2 & 3 & 4 \end{bmatrix}$$

or, equally well,

$$H = \begin{bmatrix} 0 & 1 & 2 & 1 & 2 & 4 \\ 3 & 0 & 2 & 2 & 1 & 1 \end{bmatrix}.$$

The MLCT problem for $d = 4$

The maximum length of an $[n, n - r, 4]$-code, for given r, is equal to the value of $\max_3 (r, q)$, the largest size of an $(n, 3)$-set in $V(r, q)$ (or in $PG(r - 1, q)$).

An $(n, 3)$-set in the plane $PG(2, q)$ is usually called an *n-arc*, while an $(n, 3)$-set in $PG(r - 1, q)$, for $r > 3$, is called an *n-cap*.

Since three points of $PG(r - 1, q)$ are linearly dependent if and only if they are collinear (i.e. they lie on the same line), we may describe an *n*-arc/*n*-cap as a set of n points, no three of which are collinear.

The problem of determining the values of $\max_3 (r, q)$, first considered by Bose (1947), was quickly solved for $q = 2$, for all r, and for $r \leqslant 4$, for all q. But, despite having received much attention since, the problem has been solved only for the additional pairs $(r, q) = (4, 3)$ and $(5, 3)$. The known values of $\max_3 (r, q)$ are listed in Fig. 14.9.

$$\max_3 (r, 2) = 2^{r-1} \qquad \text{(Bose 1947)}$$

$$\max_3 (3, q) = \begin{cases} q + 1, & q \text{ odd} \\ q + 2, & q \text{ even} \end{cases} \qquad \text{(Bose 1947)}$$

$$\max_3 (4, q) = \begin{cases} q^2 + 1, & q \text{ odd} \quad \text{(Bose 1947)} \\ q^2 + 1, & q \text{ even} \quad \text{(Qvist 1952)} \end{cases}$$

$$\max_3 (5, 3) = 20 \qquad \text{(Pellegrino 1970)}$$

$$\max_3 (6, 3) = 56 \qquad \text{(Hill 1973)}$$

Fig. 14.9. The known values of $\max_3 (r, q)$.

We now prove the more straightforward of these results.

The determination of $\max_3 (r, 2)$

Here we are concerned with finding optimal *binary* linear codes with $d = 4$. The following general theorem shows that we may obtain such codes from optimal codes of minimum distance 3 by the simple device of adding an overall parity-check.

Theorem 14.10 Suppose d is odd. Then there exists a binary $[n, k, d]$-code if and only if there exists a binary $[n + 1, k, d + 1]$-code.

Proof The proof of Theorem 2.7 is valid in the restriction to linear codes. This is because an 'extended' linear code (i.e. the code obtained from a linear code by adding an overall parity-check) is also linear (see Exercise 5.4).

Corollary 14.11   Suppose d is even. Then
(i) $B_2(n, d) = B_2(n - 1, d - 1)$
(ii) $\max_{d-1}(r, 2) = \max_{d-2}(r - 1, 2) + 1$.

Proof
(i)   is immediate from Theorem 14.10.
(ii) $n \leq \max_{d-1}(r, 2) \Leftrightarrow$ there exists a binary
$$[n, n - r, d]\text{-code}$$
$$\Leftrightarrow \text{there exists a binary}$$
$$[n - 1, n - r, d - 1]\text{-code}$$
$$\Leftrightarrow n - 1 \leq \max_{d-2}(r - 1, 2)$$
$$\Leftrightarrow n \leq \max_{d-2}(r - 1, 2) + 1.$$

Corollary 14.12 $\max_3(r, 2) = 2^{r-1}$.

Proof   By Theorem 14.4, $\max_2(r, 2) = 2^r - 1$. Hence
$$\max_3(r, 2) = (2^{r-1} - 1) + 1 = 2^{r-1}.$$

The optimal binary code with $d = 4$ and redundancy r is the extended Hamming code Hâm $(r - 1, 2)$. As we saw in Chapter 8, a parity-check matrix for this code is

$$\bar{H} = \begin{bmatrix} & & & 0 \\ & H & & \vdots \\ & & & 0 \\ 1 & 1 & \cdots & 1 \end{bmatrix},$$

where H is a parity-check matrix for Ham $(r - 1, 2)$, so that the columns of H are just the points of $PG(r - 2, 2)$ (i.e. the non-zero vectors of $V(r - 1, 2)$).

The columns of $\bar{H}$ form an optimal 2^{r-1}-cap in $PG(r - 1, 2)$. It consists of the points of $PG(r - 1, 2)$ *not* lying in the subspace $\{(x_1, \ldots, x_r) \mid x_r = 0\}$. Geometrically, it may be described as the complement of a hyperplane.

The determination of max₃ (3, *q*)

First we give some examples of good linear codes with $d = 4$ and redundancy 3. We then prove that these codes are optimal by showing that there cannot exist such codes of greater length.

Theorem 14.13 Let $a_1, a_2, \ldots, a_{q-1}$ be the non-zero elements of $GF(q)$.

(i) The matrix $H = \begin{bmatrix} 1 & 1 & \cdots & 1 & 1 & 0 \\ a_1 & a_2 & \cdots & a_{q-1} & 0 & 0 \\ a_1^2 & a_2^2 & \cdots & a_{q-1}^2 & 0 & 1 \end{bmatrix}$

   is the parity-check matrix of a $[q+1, q-2, 4]$-code. Equivalently, the columns of H form a $(q+1)$-arc in $PG(2, q)$.

(ii) If q is even, then the matrix

$$H^* = \begin{bmatrix} 1 & 1 & \cdots & 1 & 1 & 0 & 0 \\ a_1 & a_2 & \cdots & a_{q-1} & 0 & 1 & 0 \\ a_1^2 & a_2^2 & \cdots & a_{q-1}^2 & 0 & 0 & 1 \end{bmatrix}$$

   is the parity-check matrix of a $[q+2, q-1, 4]$-code. Equivalently, the columns of H^* form a $(q+2)$-arc in $PG(2, q)$.

Proof (i) It is enough to show that any three columns of H are linearly independent. Any three of the first $q-1$ columns of H form a Vandermonde matrix, and so are linearly independent by Theorems 11.1 and 11.2. For any three columns which include one or both of the last two columns, the determinant may be expanded about these last columns to get again the determinant of a Vandermonde matrix.

(ii) We have shown in the proof of part (i) that any three columns of H^* are linearly independent, with the possible exception of three of the form

$$\begin{bmatrix} 1 \\ a_i \\ a_i^2 \end{bmatrix}, \begin{bmatrix} 1 \\ a_j \\ a_j^2 \end{bmatrix}, \text{ and } \begin{bmatrix} 0 \\ 1 \\ 0 \end{bmatrix}.$$

The determinant of the matrix A formed by these three columns is equal to $a_i^2 - a_j^2$. Since q is even, $GF(q)$ has characteristic 2 (cf. Exercise 4.6). Hence, by Exercise 3.12, $a_i^2 - a_j^2 = (a_i - a_j)^2$. Since $a_i \neq a_j$, $\det A$ is non-zero.

Corollary 14.14 $\quad \max_3 (3, q) \geq \begin{cases} q + 1, & \text{if } q \text{ is odd} \\ q + 2, & \text{if } q \text{ is even.} \end{cases}$

Remark The $(q + 1)$-arc formed by the columns of H in Theorem 14.13 is the *conic* $\{(x, y, z) \in PG(2, q) \mid yz = x^2\}$.

We now show that the codes/arcs given in Theorem 14.13 are optimal.

Theorem 14.15
(i) For any prime power q, $\max_3 (3, q) \leq q + 2$.
(ii) If q is odd, then $\max_3 (3, q) \leq q + 1$.

First proof (i) Let H be a standard form parity-check matrix for an $[n, n - 3, 4]$-code C over $GF(q)$, with $n = \max_3 (3, q)$:

$$H = \begin{bmatrix} a_1 & a_2 & \cdots & a_{n-3} & 1 & 0 & 0 \\ b_1 & b_2 & \cdots & b_{n-3} & 0 & 1 & 0 \\ c_1 & c_2 & \cdots & c_{n-3} & 0 & 0 & 1 \end{bmatrix}.$$

Since any three columns of H are linearly independent, the determinant formed by any three columns must be non-zero. From the non-vanishing of the determinant formed by any two of the last three columns and one of the first $n - 3$ columns, we find that the a_is, b_is and c_is are all non-zero. Multiplying the ith column by a_i^{-1} for $i = 1, 2, \ldots, n - 3$, we have that C is equivalent to a code in which the a_is are all 1. Thus we may assume that

$$A = \begin{bmatrix} 1 & 1 & \cdots & 1 & 1 & 0 & 0 \\ b_1 & b_2 & \cdots & b_{n-3} & 0 & 1 & 0 \\ c_1 & c_2 & \cdots & c_{n-3} & 0 & 0 & 1 \end{bmatrix},$$

where the b_is and c_is are non-zero. As the determinant formed by the last column and two of the first $n - 3$ columns is non-zero, the b_is must be distinct non-zero elements of $GF(q)$. Hence $n - 3 \leq q - 1$ and so $n \leq q + 2$.

(ii) (Adapted from Fenton and Vámos, 1982). Now suppose q is odd. Suppose, for a contradiction, that a $[q + 2, q - 1, 4]$-code C exists. Then, as in (i), we may assume that C has a parity-check matrix

$$H = \begin{bmatrix} 1 & 1 & \cdots & 1 & 1 & 0 & 0 \\ b_1 & b_2 & \cdots & b_{q-1} & 0 & 1 & 0 \\ c_1 & c_2 & \cdots & c_{q-1} & 0 & 0 & 1 \end{bmatrix}$$

where $b_1, b_2, \ldots, b_{q-1}$ are the distinct non-zero elements of $GF(q)$ and similarly $c_1, c_2, \ldots, c_{q-1}$ are also the distinct non-zero elements of $GF(q)$ in some order. The non-vanishing of determinants of the form

$$\det \begin{bmatrix} 1 & 1 & 1 \\ b_i & b_j & 0 \\ c_i & c_j & 0 \end{bmatrix}$$

implies that the elements $b_1 c_1^{-1}, b_2 c_2^{-1}, \ldots, b_{q-1} c_{q-1}^{-1}$ are distinct and so they too are the non-zero elements of $GF(q)$ in some order. Hence, by Exercise 3.13, all three of the products $\prod_{i=1}^{q-1} b_i$, $\prod_{i=1}^{q-1} c_i$, and $\prod_{i=1}^{q-1} (b_i c_i^{-1})$ are equal to -1. But then

$$\prod_{i=1}^{q-1} (b_i c_i^{-1}) = \left(\prod_{i=1}^{q-1} b_i \right) \left(\prod_{i=1}^{q-1} c_i \right)^{-1} = (-1)(-1)^{-1} = 1.$$

Since $1 \neq -1$ if q is odd, this gives the desired contradiction.

Second proof (geometric) (i) Suppose K is an n-arc in $PG(2, q)$ of maximum size $n = \max_3 (3, q)$. Let P be a point of K. By Lemma 14.6(iv), there are $q + 1$ lines through P and every point of K lies on one or other of them. But on none of these lines can there be more than one point of K besides P (by definition of an n-arc, no three points of K are collinear). Thus $n \leq 1 + (q + 1) = q + 2$.

(ii) Now suppose q is odd. Suppose, for a contradiction, that K is a $(q + 2)$-arc in $PG(2, q)$. Then if P is any point of K, each of the $q + 1$ lines through P must contain exactly one further point of K. This means that every line in $PG(2, q)$ meets K in either 2 or 0 points (but never in 1). Now let Q be any point of $PG(2, q)$ lying outside K. Through Q there pass $q + 1$ lines and each point of K lies on one (and only one) of them. So if t of these lines meet K in two points, then $|K| = 2t$, contradicting $|K| = q + 2$ being odd.

Remark The author feels that the attractiveness of the above proofs merits the inclusion of both. The geometric proof has two important advantages: (1) it generalizes to give upper bounds on $\max_3(r, q)$ for larger values of r; (2) it does not assume specific properties of the field $GF(q)$, and so gives the same upper bound on the size of n-arcs in any projective plane of order q.

Corollary 14.14 and Theorem 14.15 give

Theorem 14.16   (Bose 1947)

$$\max_3(3, q) = \begin{cases} q + 1 & \text{if } q \text{ is odd} \\ q + 2 & \text{if } q \text{ is even.} \end{cases}$$

Remark It has been shown by Segre (1954) that, for q odd, every $(q + 1)$-arc in $PG(2, q)$ is a conic. This implies that the optimal $[q + 1, q - 2, 4]$-code is unique, up to equivalence. For q even, optimal $(q + 2)$-arcs in $PG(2, q)$ are not in general unique, and a classification is unknown.

The determination of $\max_3(4, q)$, for q odd

As we shall adopt a geometric approach here, we introduce a little more terminology concerning the projective geometry $PG(r - 1, q)$. In defining $PG(r - 1, q)$ from the vector space $V(r, q)$, recall that the points and lines in $PG(r - 1, q)$ are the 1- and 2-dimensional subspaces respectively of $V(r, q)$. More generally we define a *t-space* in $PG(r - 1, q)$ to be a $(t + 1)$-dimensional subspace of $V(r, q)$. Thus a 0-space is a point and a 1-space is a line. A 2-space is called a *plane* and an $(r - 2)$-space in $PG(r - 1, q)$ is called a *hyperplane*. Note that the *dimension t* of a *t*-space in $PG(r - 1, q)$ is always one less than the corresponding vector space dimension.

We usually identify a *t*-space in $PG(r - 1, q)$ with the set of points it contains. The number of points in a *t*-space is $(q^{t+1} - 1)/(q - 1)$, since a $(t + 1)$-dimensional subspace of $V(r, q)$ contains $q^{t+1} - 1$ non-zero vectors, each of which has $q - 1$

non-zero scalar multiples. A *t*-space is just a copy of $PG(t, q)$ in so far as the incidence properties of its subspaces are concerned. In particular, a cap in $PG(r - 1, q)$ must meet a $(t - 1)$-space in at most $\max_3 (t, q)$ points, bearing in mind that any subset of a cap is also a cap.

We may now derive an upper bound on $\max_3 (4, q)$, for q odd.

Theorem 14.17 If q is odd, then $\max_3 (4, q) \leq q^2 + 1$.

Proof Suppose K is an n-cap in $PG(3, q)$ of maximum size. Let P_1 and P_2 be any two points of K and let L be the line on which P_1 and P_2 lie. Since no three points of K are collinear, L contains no other point of K. Through the line L there pass $q + 1$ planes (Exercise 14.4), and each point of K, other than P_1 and P_2, lies on one and only one of these planes. Since q is odd, it follows from Theorem 14.15(ii) that no plane can contain more than $q + 1$ points of K. In particular, a plane through L can contain at most $q - 1$ points in addition to P_1 and P_2. Hence

$$n \leq 2 + (q + 1)(q - 1) = q^2 + 1.$$

We next show that $(q^2 + 1)$-caps exist in $PG(3, q)$, when q is odd.

Theorem 14.18 Suppose q is odd and let b be a non-square in $GF(q)$. Then the set

$$Q = \{(x, y, z, w) \in PG(3, q) \mid zw = x^2 - by^2\}$$

is a $(q^2 + 1)$-cap in $PG(3, q)$.

Proof Since b is a non-square, the only point of Q having $z = 0$ is $(0, 0, 0, 1)$. Each of the remaining points may be represented by a vector having $z = 1$, and so we may write

$$Q = \{(0, 0, 0, 1), (x, y, 1, x^2 - by^2) \mid (x, y) \in V(2, q)\}. \quad (14.19)$$

This shows that $|Q| = q^2 + 1$. We must show that no three points of Q are collinear. Clearly $(0, 0, 0, 1)$ cannot be collinear with two other points of Q because there is only one point of Q of the form $(x, y, 1, *)$ for any given pair (x, y). Now let $\mathbf{a}_1 = (x_1, y_1, 1, x_1^2 - by_1^2)$ and $\mathbf{a}_2 = (x_2, y_2, 1, x_2^2 - by_2^2)$ be any two points of Q, other than $(0, 0, 0, 1)$. Suppose, for a contradiction, that the line joining $\mathbf{a}_1$ and $\mathbf{a}_2$ contains a third point of Q. Then, for some non-zero scalar λ, $\mathbf{a}_1 + \lambda\mathbf{a}_2 \in Q$, i.e. the point $(x, y, z, w) = (x_1 + \lambda x_2, y_1 + \lambda y_2, 1 + \lambda, x_1^2 - by_1^2 + \lambda x_2^2 - \lambda by_2^2)$ satisfies $zw = x^2 - by^2$. This condition implies, after some cancellation, that

$$\lambda x_1^2 + \lambda x_2^2 - \lambda by_1^2 - \lambda by_2^2 = 2\lambda x_1 x_2 - 2\lambda by_1 y_2.$$

Since $\lambda \neq 0$, it follows that

$$(x_1 - x_2)^2 = b(y_1 - y_2)^2,$$

which is impossible since b is a non-square.

Putting Theorems 14.17 and 14.18 together gives

Theorem 14.20 If q is odd, then $\max_3 (4, q) = q^2 + 1$.

Example 14.21 Take $q = 3$ and $b = -1$ in Theorem 14.18. By (14.19), a 10-cap in $PG(3, 3)$ is formed by the columns of the matrix

$$H = \begin{bmatrix} 0 & 0 & 0 & 0 & 1 & 1 & 1 & 2 & 2 & 2 \\ 0 & 0 & 1 & 2 & 0 & 1 & 2 & 0 & 1 & 2 \\ 0 & 1 & 1 & 1 & 1 & 1 & 1 & 1 & 1 & 1 \\ 1 & 0 & 1 & 1 & 1 & 2 & 2 & 1 & 2 & 2 \end{bmatrix}.$$

Thus H is the parity-check matrix of a ternary $[10, 6, 4]$-code which is of greatest length for $d = 4$ and $r = 4$.

Remarks (1) The set Q of Theorem 14.18 is an example of an *elliptic quadric*. For q odd, any elliptic quadric is a $(q^2 + 1)$-cap, and conversely (Barlotti 1955) any $(q^2 + 1)$-cap is an elliptic quadric. This implies that the optimal $[q^2 + 1, q^2 - 3, 4]$-code is unique, up to equivalence.

(2) For $q = 2^h$, with $h > 1$, it is also true that $\max_3 (4, q) = q^2 + 1$, but the proof is a little trickier and is omitted here.

The values of $B_q(n, 4)$, for $n \leq q^2 + 1$

By means of Theorem 14.3, we can instantly translate our results concerning $\max_3 (r, q)$ for $r = 2$ and 3 into results about $B_q(n, 4)$.

Theorem 14.22 If q is odd, then

$$B_q(n, 4) = \begin{cases} q^{n-3} & \text{for } 4 \leq n \leq q + 1 \\ q^{n-4} & \text{for } q + 2 \leq n \leq q^2 + 1. \end{cases}$$

If q is even, then

$$B_q(n, 4) = \begin{cases} q^{n-3} & \text{for } 4 \leqslant n \leqslant q + 2 \\ q^{n-4} & \text{for } q + 3 \leqslant n \leqslant q^2 + 1. \end{cases}$$

Remarks on max$_3$ (r, q) for $r \geqslant 5$

For $r = 3$ and $r = 4$ the packing problem for caps in $PG(r - 1, q)$ was fairly easy to solve because of the existence of natural geometric configurations (conics in $PG(2, q)$ and elliptic quadrics in $PG(3, q)$) which are optimal caps. But in $PG(r - 1, q)$ for $r \geqslant 5$, large caps do not appear to arise in such a natural way and so the packing problem is much more difficult. As we see from Table 14.9, the only known values of max$_3$ (r, q) for $q \neq 2$ and $r \geqslant 5$ are max$_3$ $(5, 3) = 20$ and max$_3$ $(6, 3) = 56$. (For a coding-theoretic proof of the second result, wherein the uniqueness of the optimal ternary $[56, 50, 4]$-code is also demonstrated, see Hill (1978).)

It is easy to construct 20-caps in $PG(4, 3)$ (Exercise 14.9) but hard to show that 20 is the largest size possible. By contrast, it is rather difficult to describe a 56-cap in $PG(5, 3)$, but a short proof of the maximality of 56 has been given by Bruen and Hirschfeld (1978) (cf. Exercise 14.11). In the next dimension up for $q = 3$, the best known bounds are

$$112 \leqslant \max{}_3 (7, 3) \leqslant 163,$$

suggesting that the problem of finding optimal caps in $PG(6, 3)$ is far from solution.

Concluding remarks on Chapter 14

(1) We have mentioned that the problem of determining max$_s$ (r, q) was first considered by Bose (1947). Much of the subsequent work has been carried out by the Italian school of geometers led by Segre, Barlotti and Tallini.

For a survey of the known results concerning max$_s$ (r, q) and similar functions, see Hirschfeld (1983). For a comprehensive coverage of the theory of projective geometries over finite fields, see Hirschfeld (1979 and Volume 2, to appear).

(2) For recent results concerning max$_s$ (r, q) for $q = 3$ and $s \leqslant r \leqslant 15$, see Games (1983).

(3) There seems to be little pattern to results concerning $\max_{d-1}(r, q)$ for fixed values of d greater than 4. However, when d takes its maximum value for given r, that is $d = r + 1$, an interesting pattern once again emerges. This case is the subject of the next chapter.

(4) Another version of the MLCT problem is to find, for given q, n and k, the maximum value of d for which there exists an $[n, k, d]$-code over $GF(q)$. In the case of binary linear codes, Helgert and Stinaff (1973) give a table of such values (or bounds when the values are not known) for $k \le n \le 127$. For a comprehensive update of this table, incorporating many improvements by various authors, see Verhoeff (1985).

Exercises 14

14.1   Is it true that $B_2(n, d)$ is always equal to the highest power of 2 less than or equal to $A_2(n, d)$?

14.2   Show that (i) $B_q(n, 1) = q^n$, (ii) $B_q(n, 2) = q^{n-1}$.

14.3   Show that $B_q(n, 3) = q^{\lfloor n - \log_q(nq - n + 1) \rfloor}$.

14.4   Show that in $PG(3, q)$ the number of planes containing a given line is $q + 1$.

14.5   Which code is the optimal $[n, n - 5, 5]$-code having $n = \max_4(5, 3)$?

14.6   Specify a $[26, 22, 4]$-code over $GF(5)$.

14.7   Pinpoint where the proofs of Theorems 14.17 and 14.18 fail when q is even.

14.8   Devise a syndrome-decoding algorithm for a $[q^2 + 1, q^2 - 3, 4]$-code over $GF(q)$ (q odd), which will correct any single error and detect any double error.

14.9   Given the 10-cap of Example 14.21, construct a 20-cap in $PG(4, 3)$.

14.10 Show that, in $PG(m, q)$, the number of $(t + 1)$-spaces containing a given t-space is $(q^{m-t} - 1)/(q - 1)$. In $PG(5, 3)$, state (i) how many planes contain a given line, (ii) how many 3-spaces contain a given plane, (iii) how many 4-spaces contain a given 3-space.

14.11 Given that $\max_3(5, 3) = 20$, show that $\max_3(6, 3) \le 56$. [*Hint*: Use parts (i), (ii) and (iii) of Exercise 14.10.]

14.12 State the values of $B_3(n, 4)$ for $4 \le n \le 112$.

15 MDS codes

In the previous chapter we considered the problem of finding linear codes of maximum length for given redundancy r and given minimum distance d. Particular attention was paid to the cases $d \leq 4$. In this chapter we consider the problem when d is as large as possible for given redundancy r. The following theorem shows that this is the case $d = r + 1$.

Theorem 15.1 An $[n, n - r, d]$-code satisfies $d \leq r + 1$.

Proof 1 This is just the Singleton bound applied to linear codes. Theorem 10.17 states that any q-ary (n, M, d)-code satisfies $M \leq q^{n-d+1}$. So, in particular, an $[n, n - r, d]$-code over $GF(q)$ satisfies $q^{n-r} \leq q^{n-d+1}$, whence $d \leq r + 1$.

Proof 2 Suppose C is an $[n, n - r, d]$-code and let $G = [I_{n-r} \mid A]$ be a standard form generator matrix of C. Since A has r columns, those codewords which are rows of G have weight $\leq r + 1$. The result follows by Theorem 5.2.

Definition An $[n, n - r, r + 1]$-code (i.e. a linear code of redundancy r whose minimum distance is equal to $r + 1$) is called a *maximum distance separable code*, or *MDS code* for short.

By Theorem 14.1, the maximum length of an $[n, n - r, r + 1]$-code over $GF(q)$ is equal to the value of $\max_r (r, q)$, the largest size of an (n, r)-set in $V(r, q)$. We recall that an (n, r)-set in $V(r, q)$ is a set of n vectors such that any r of them are linearly independent. Equivalently, an (n, r)-set in $V(r, q)$ is a set of n vectors such that any r of them form a basis for $V(r, q)$.

MDS codes were first studied explicitly by Singleton (1964), although the problem of finding $\max_r (r, q)$ had already been studied as a problem in statistics (Bush 1952) and as a problem in geometry (Segre 1955, 1961). (In the geometrical context, an

(n, r)-set, regarded as a subset of $PG(r-1, q)$, is called an *n-arc*. This agrees with the usage of the term *n-arc* for an $(n, 3)$-set in $PG(2, q)$ already met in Chapter 14.)

MacWilliams and Sloane (1977) introduce their chapter on MDS codes as 'one of the most fascinating in all of coding theory'. The problem of determining the values of $\max_r (r, q)$ is a particularly attractive one for two reasons. Firstly, the problem is equivalent to a surprising list of combinatorial problems; no fewer than six different interpretations are given in MacWilliams' and Sloane's book, while yet another is given in Fenton and Vámos (1982). Secondly, although a complete solution to the problem seems inaccessible at present, the known results suggest a tantalizingly simply stated conjecture:

Conjecture 15.2 If $2 \le r \le q$, then

$$\max_r (r, q) = q + 1$$

(except that $\max_3 (3, q) = \max_{q-1} (q - 1, q) = q + 2$ if $q = 2^h$).

Note that the conjecture has already been proved for $r = 2$ (Theorem 14.4) and for $r = 3$ (Theorem 14.16). Before considering the conjecture further let us dispose of the rather uninteresting cases outside the range to which it applies. For redundancies 0 and 1, MDS codes exist of any length n over any field $GF(q)$ (for $r = 0$, $V(n, q)$ is an $[n, n, 1]$-code, while for $r = 1$, the matrix

$$\left[I_{n-1} \ \begin{array}{|c} 1 \\ \vdots \\ 1 \end{array} \right]$$

generates an $[n, n-1, 2]$-code). Cases $r > q$ are covered by the following theorem.

Theorem 15.3 If $r \ge q$, then $\max_r (r, q) = r + 1$. Any MDS code of redundancy $r \ge q$ is equivalent to a repetition code of length $r + 1$.

Proof The repetition code of length $r + 1$ is an $[r + 1, 1, r + 1]$-code with generator matrix $[1\ 1\ \cdots\ 1]$. Hence

$$\max_r (r, q) \ge r + 1.$$

Also, it is clear that any $[r+1, 1, r+1]$-code is equivalent to a repetition code.

Now suppose $r \geq q$ and suppose for a contradiction that $\max_r (r, q) \geq r + 2$. Then there exists an $[r+2, 2, r+1]$-code C over $GF(q)$. This code C must be equivalent to a code having generator matrix

$$G = \begin{bmatrix} 1 & 0 & 1 & 1 & \cdots & 1 \\ 0 & 1 & a_1 & a_2 & \cdots & a_r \end{bmatrix}.$$

In order that any linear combination of the rows of G has weight at least $r+1$, the a_is must be distinct non-zero elements of $GF(q)$. This implies that $r \leq q - 1$, contrary to hypothesis.

Remark It follows from Theorem 15.3 and the preceding remarks that the only *binary* MDS codes are $V(n, 2)$, the even weight codes E_n, and repetition codes. So this chapter is really of interest only for codes over $GF(q)$ with $q > 2$.

From now on we assume that r lies in the range $2 \leq r \leq q$ and return to our consideration of Conjecture 15.2. Our first task will be to show that there exist MDS codes which meet the conjectured values of $\max_r (r, q)$ in all cases.

Theorem 15.4 Suppose $2 \leq r \leq q$. Let $a_1, a_2, \ldots, a_{q-1}$ be the non-zero elements of $GF(q)$. Then the matrix

$$H = \begin{bmatrix} 1 & 1 & \cdots & 1 & 1 & 0 \\ a_1 & a_2 & \cdots & a_{q-1} & 0 & 0 \\ a_1^2 & a_2^2 & \cdots & a_{q-1}^2 & 0 & 0 \\ \vdots & \vdots & & \vdots & \vdots & \vdots \\ & & & & 0 & 0 \\ a_1^{r-1} & a_2^{r-1} & \cdots & a_{q-1}^{r-1} & 0 & 1 \end{bmatrix}$$

is the parity-check matrix of an MDS $[q+1, q+1-r, r+1]$-code. Equivalently, the columns of H form a $(q+1)$-arc in $PG(r-1, q)$.

Proof This is exactly the same as the proof of Theorem 14.13(i), for the determinant of a matrix formed by any r columns of H is equal to the determinant of a Vandermonde matrix and so is non-zero. Thus any r columns of H are linearly independent.

Corollary 15.5 If $2 \leq r \leq q$, then $\max_r (r, q) \geq q + 1$.

As we saw in Theorem 14.13(ii), in the case where q is even and $r = 3$, we may add the further column

$$\begin{bmatrix} 0 \\ 1 \\ 0 \end{bmatrix}$$

to the matrix H of Theorem 15.4 to get an MDS code of length $q + 2$. Such a trick will not work for $r > 3$. However, we see from Conjecture 15.2 that the case q even and $r = q - 1$ also seems to be special. Indeed there exists an MDS code of length $q + 2$ in this case too. This fact will follow from the very useful result that the dual code of an MDS code is also MDS, thus implying that the roles of dimension and redundancy are interchangeable in so far as the existence of MDS codes is concerned. In order to show this duality, we first reformulate our problem in terms of matrices having every square submatrix non-singular.

Definitions A square matrix is called *non-singular* if its columns are linearly independent, or equivalently, if it has a non-zero determinant (cf. Theorem 11.2).

Given any matrix A, a $t \times t$ *square submatrix* of A is a $t \times t$ matrix consisting of the entries of A lying in some t rows and some t columns of A.

For example, if

$$A = \begin{bmatrix} a_{11} & a_{12} & a_{13} & a_{14} \\ a_{21} & a_{22} & a_{23} & a_{24} \\ a_{31} & a_{32} & a_{33} & a_{34} \end{bmatrix},$$

then

$$\begin{bmatrix} a_{22} & a_{23} \\ a_{32} & a_{33} \end{bmatrix} \quad \text{and} \quad \begin{bmatrix} a_{11} & a_{14} \\ a_{31} & a_{34} \end{bmatrix}$$

are examples of 2×2 square submatrices of A.

Theorem 15.6 Suppose C is an $[n, n - r]$-code with parity-check matrix $H = [A^T \mid I_r]$. Then C is an MDS code (i.e. $d(C) = r + 1$) if and only if every square submatrix of A is non-singular.

Proof By Theorem 8.4, C is an MDS code if and only if any r columns of H are linearly independent, i.e. if and only if any $r \times r$ submatrix of H is non-singular. Let us interpret this last condition on H as a condition on A^T. Suppose B is an $r \times r$ submatrix of H obtained by choosing some r columns of H. Suppose t of the chosen columns are from A^T and $r - t$ of them from I_r. If we expand $\det B$ about the last $r - t$ columns, we end up with

$$\det B = \pm \det B',$$

where B' is the $t \times t$ matrix obtained by taking the $r \times t$ matrix consisting of the t chosen columns of A^T and then deleting the $r - t$ rows corresponding to where the chosen columns of I_r have 1s. To illustrate this point suppose

$$H = \begin{bmatrix} a_{11} & a_{21} & a_{31} & 1 & 0 & 0 & 0 \\ a_{12} & a_{22} & a_{32} & 0 & 1 & 0 & 0 \\ a_{13} & a_{23} & a_{33} & 0 & 0 & 1 & 0 \\ a_{14} & a_{24} & a_{34} & 0 & 0 & 0 & 1 \end{bmatrix}.$$

If B is the 4×4 submatrix of H consisting of columns 1, 3, 5 and 6, then

$$\det B = \det \begin{bmatrix} a_{11} & a_{31} & 0 & 0 \\ a_{12} & a_{32} & 1 & 0 \\ a_{13} & a_{33} & 0 & 1 \\ a_{14} & a_{34} & 0 & 0 \end{bmatrix} = \det \begin{bmatrix} a_{11} & a_{31} \\ a_{14} & a_{34} \end{bmatrix} = \det B'.$$

Returning to the general case, it follows that B is non-singular if and only if the corresponding square submatrix B' is non-singular. It is clear that *any* $t \times t$ square submatrix B' of A^T (for any t with $1 \leqslant t \leqslant r$) arises from some $r \times r$ submatrix B of H in this way, and so the result follows.

Corollary 15.7 The dual code of an MDS code is also MDS.

Proof The code C with parity-check matrix $[A^T \mid I_r]$ is MDS $\Leftrightarrow A^T$ has the property that every square submatrix is non-singular

$\Leftrightarrow A$ has the same property (since the determinant of any square matrix is equal to the determinant of its transpose)

$\Leftrightarrow$ the code $C^{\perp}$ with parity-check matrix $[I_{n-r} \mid -A]$ is MDS.

It follows from Corollary 15.7 that generator matrices and parity-check matrices of MDS $[n, k]$-codes serve also as parity-check matrices and generator matrices respectively of MDS $[n, n-k]$-codes.

Corollary 15.8   There exists an MDS $[n, k]$-code over $GF(q)$ if and only if there exists an MDS $[n, n-k]$-code over $GF(q)$.

Corollary 15.9   Suppose $q = 2^h$, $h > 1$. Then there exists a $[q+2, 3, q]$-code over $GF(q)$. Equivalently, there exists a $(q+2)$-arc in $PG(q-2, q)$.

Proof   By Theorem 14.13(ii), there exists a $[q+2, q-1, 4]$-code over $GF(q)$. By Corollary 15.7, its dual code is a $[q+2, 3, q]$-code.

Combining the results of Corollaries 14.14, 15.5 and 15.9, we have

Theorem 15.10   If $2 \le r \le q$, then $\max_r(r, q) \ge q + 1$. If also $q = 2^h$ and $r = 3$ or $q - 1$, then $\max_r(r, q) \ge q + 2$.

The known results concerning Conjecture 15.2

Theorem 15.10 shows that the conjectured values of $\max_r(r, q)$ are all lower bounds. The conjecture was shown to be true for $r = 2$ and $r = 3$ in Theorems 14.4 and 14.16. We mention without proof that, by geometric methods, the conjecture has also been proved for $r = 4$ and $r = 5$, for all q (Segre 1955 and Casse 1969). Using the duality result of Corollary 15.8, the truth of the conjecture for $r \le 5$ implies its truth also for r in the range $q - 3 \le r \le q$ (see Exercises 15.2 and 15.3). [This last result was first proved in a different way by Thas (1968), who also showed (1968, 1969) that the conjecture is true for q odd in the ranges $q > (4r - 9)^2$ and $q - 3 > r > q - \frac{1}{4}\sqrt{q} - 5/4$].

Following MacWilliams and Sloane (1977), we show the results graphically in Fig. 15.11, which neatly illustrates the symmetry between dimension k and redundancy r.

The broken line $n = k + r = q + 1$ in Fig. 15.11 is the conjectured bound above which no MDS code is known to exist. The heavy line represents an upper bound given by repeated application of the recursive bound

$$\max_{r+1} (r + 1, q) \leqslant \max_r (r, q) + 1$$

(see Exercise 15.5), starting at $\max_5 (5, q) = q + 1$ (thus $\max_6 (6, q) \leqslant q + 2$, $\max_7 (7, q) \leqslant q + 3, \ldots, \max_r (r, q) \leqslant q + r - 4$ for $r \geqslant 6$). The region marked with a question mark is therefore the 'grey' area where the existence of MDS codes is undecided.

Finally we mention that the conjecture has been verified by exhaustive search for $q \leqslant 11$, for all r (Maneri and Silverman

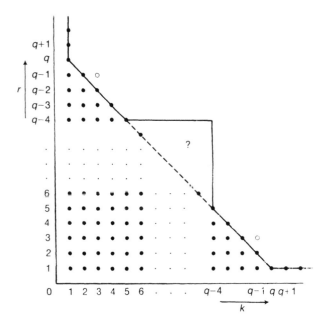

Fig. 15.11. Values of k, r for which a $[k + r, k]$-MDS code exists. ● means MDS code exists for all q. ○ means MDS code exists if and only if $q = 2^h$.

1966 and Jurick 1968) and so the smallest undecided case is $\max_6 (6, 13) = 14$ or 15.

Concluding remarks on Chapter 15

(1) The Reed–Solomon codes described in Chapter 11 are MDS codes; they are shortened versions of the codes defined in Theorem 15.4. Since MDS codes meet the Singleton bound, Theorem 15.4 enables Theorem 11.4 to be improved to

Theorem 15.12   If q is a prime power and if $d \leq n \leq q + 1$, then

$$A_q(n, d) = B_q(n, d) = q^{n-d+1}.$$

(2) One remarkable property of an MDS $[n, k]$-code C over $GF(q)$ is that its weight enumerator is completely determined by the values of n, k and q and does not depend on the code C itself. This fact is a little less surprising when one considers the MacWilliams identity. Let

$$W_C(z) = 1 + \sum_{i=n-k+1}^{n} A_i z^i$$

be the weight enumerator of C. Since $C^\perp$ is also MDS and hence has minimum distance $k + 1$, the coefficients of $z, z^2, \ldots, z^k$ on the right-hand side of the MacWilliams identity (Theorem 13.6) must all be equal to zero, giving k equations in the k unknowns $A_{n-k+1}, \ldots, A_n$ (we also have the equation $1 + \sum_{i=n-k+1}^{n} A_i = q^k$). It turns out that these equations have a unique solution. Exercise 15.6 gives an illustration of this. In fact it is possible to derive the formulae

$$A_i = \binom{n}{i}(q - 1) \sum_{j=0}^{i-d} (-1)^j \binom{i-1}{j} q^{i-d-j} \tag{15.13}$$

for the A_is in terms of n, d and q, though this derivation is a little complicated (see e.g. Chapter 11 of MacWilliams and Sloane, 1977) and is not included here.

(3) Theorem 15.6 enables the MDS codes existence problem to be posed in elementary terms, independently of any terminology from coding theory or geometry. In view of Theorem 15.10, Conjecture 15.2 may be simply stated as follows.

Conjecture 15.14 Any $r \times k$ matrix over $GF(q)$ with $2 \leq r$, $k \leq q$ and having the property that any square sub-matrix is non-singular satisfies

$$r + k \leq q + 1$$

(except for case $q = 2^h$ and r or $k = 3$).

From an earlier remark, the smallest possible counter-example is a 6×9 or 7×8 matrix over $GF(13)$.

Exercises 15

15.1 Consider the matrix

$$A = \begin{bmatrix} 1 & 6 & 2 & 5 & 1 \\ 1 & 4 & 3 & 3 & 6 \\ 1 & 5 & 5 & 1 & 5 \end{bmatrix}$$

over $GF(7)$. Check that every square submatrix of A is non-singular. Hence write down generator matrices for $[8, 3]$ and $[8, 5]$ MDS codes over $GF(7)$.

15.2 Show that if $\max_r (r, q) = q + 1$, then $\max_{q+2-r} (q + 2 - r, q) = q + 1$.

15.3 Suppose $q = 2^h$, $h > 1$. Assuming known results about $\max_r (r, q)$ for $r \leq 4$, show that $\max_{q-1} (q - 1, q) = q + 2$.

15.4 Given that $GF(8) = \{0, a_1 = 1, a_2, a_3, \ldots, a_7\}$, write down a parity-check matrix for an $[n, n - 7, 8]$-code over $GF(8)$ with $n = \max_7 (7, 8)$.

15.5 Prove that $\max_{r+1} (r + 1, q) \leq \max_r (r, q) + 1$.

15.6 Use Theorem 13.6 to find (a) the weight enumerator of an $[8, 3, 6]$-code over $GF(7)$ and (b) the weight enumerator of an $[8, 5, 4]$-code over $GF(7)$. Check your answers by using the formulae (15.13).

15.7 For each integer $k \geq 2$, specify an $[n, k, n - k + 1]$-code over $GF(11)$ having n as large as possible.

Concluding remarks, related topics, and further reading

The main aims of this final chapter are to review the progress made in the earlier chapters and to mention some related topics, with suggestions for further reading.

The treatment presented in the book has been motivated mainly by two recurring themes:

(1) the problem of finding codes which are optimal in some sense;

(2) the problem of decoding such codes efficiently.

This has led to a rich interplay with several well-established branches of mathematics, notably algebra, combinatorics, and geometry.

With regard to optimal codes, the main emphasis has been on finding values of $A_q(n, d)$, the largest size M of an (n, M, d)-code over an alphabet of q letters. In the case of binary codes, we gave in Table 2.4 the state of knowledge regarding values of $A_2(n, d)$ for small n and d. We now consider this table again (Table 16.1), for $d \le 5$, in order to indicate those places in the text where results have been proved.

Remarks 16.2 (1) All of the bounds in Table 16.1 have been proved in the text or exercises with the exceptions of (i) the upper bounds obtained by linear programming methods and (ii) the lower bounds for $d = 3$ and $n = 9$, 10, and 11. A rather complicated construction of an $(11, 144, 3)$-code was given by Golay (1954). Successive shortenings of this code give codes with parameters $(10, 72, 3)$, $(9, 38, 3)$ and $(8, 20, 3)$. For a long time it was believed that the $(9, 38, 3)$-code was optimal, but recently Best (1980) found a $(9, 40, 3)$-code (despite a publication of 1959 which claimed that 39 was an upper bound on $A_2(9, 3)$!).

(2) It is conjectured that the Plotkin bound is always attained in the range $d \le n \le 2d + 1$. Indeed it has been shown by Levenshtein (1964) that there exist codes which meet the Plotkin

Table 16.1

Values of $A_2(n, d)$

n	$d = 3$			$d = 5$		
3	R	2	P	—		
4	R^+	2	P	—		
5	SH	4	P	R	2	P
6	SH	8	P	R^+	2	P
7	H	16	S	R^{++}	2	P
8	G	20	L_1	E_2	4	P
9	B	40	E_3	SD	6	P
10	G	72–79	L_2	SD	12	P
11	G	144–158	E_3	D	24	P
12	SH	256	L_1	SNR	32	L_1
13	SH	512	E_3	SNR	64	E_3
14	SH	1024	E_3	SNR	128	E_3
15	H	2048	S	NR	256	E_3
16	E_1	2560–3276	L_1	NR^+	256–340	L_1

Key to Table 16.1

Lower Bounds
If C is a given code, then:
C^+ denotes the code obtained from C by adding an extra zero coordinate,
SC denotes a code obtained by shortening C, possibly more than once, i.e. use E_3 (below) in the form $A_2(n-1, d) \geq \frac{1}{2}A_2(n, d)$.
R: repetition code (Example 1.11).
H: Hamming code (Theorem 8.2).
B: Best (1980).
G: Golay (1954); for alternative constructions see MacWilliams and Sloane (1977, Chapter 2, §7). A $(20, 8, 3)$-code is also constructed in Exercise 2.16.
E_1: a $(\mathbf{u} \mid \mathbf{u} + \mathbf{v})$-construction (see Exercise 2.18).
E_2: see Exercise 2.8.
D: constructed from a Hadamard design (Exercise 2.12).
NR: Nordstrom–Robinson code (Exercise 9.9).

Upper Bounds
P: Plotkin bound (Exercise 2.22).

Key to Table 16.1 (Contd.)

S: sphere-packing bound (Theorem 2.16).
L: linear programming bound (L_1: see Best *et al.* (1978) or MacWilliams and Sloane (1977); L_2: see Best (1980)).
E_3: $A_2(n, d) \leq 2A_2(n-1, d)$ (Exercise 2.2).

---

bound provided certain Hadamard matrices of order $m \leq n$ exist, for $m \equiv 0 \pmod 4$. [A *Hadamard matrix* of order m is an $m \times m$ matrix of $+1$s and -1s such that $HH^T = mI$ (over the field of real numbers). It is easy to associate a Hadamard design with a Hadamard matrix and we have already seen how such designs give rise to optimal codes (see Exercises 2.15 and 2.24)]. An introduction to Hadamard configurations may be found in Anderson (1974). A proof of Levenshtein's theorem may be found in Chapter 2 of MacWilliams and Sloane (1977). It is also a well-known conjecture that Hadamard matrices of order m exist for all positive integers $m \equiv 0 \pmod 4$. This conjecture is known to be true for $m \leq 264$ and so the Plotkin bound is indeed tight for $n \leq 264$ (in the range $2d + 1 \geq n$).

(3) Values of $A_2(n, d)$ found in the text but outside the range of Table 16.1 include:

$$A_2(23, 7) = 4096 \text{ (Theorem 11.3 or 12.20)}$$
$$A_2(n, 3) = 2^{n-r}, \text{ whenever } n = 2^r - 1 \text{ (Corollary 8.7).}$$

As well as considering optimal binary codes, much attention has also been given in this text to optimal q-ary codes for general q. For example: in Chapter 8 we showed that, for a prime power q, $A_q(n, 3) = q^{n-r}$ for any n of the form $(q^r - 1)/(q - 1)$; in Chapter 9 we showed that $A_3(11, 5) = 3^6$; in Chapter 10 we found the values of $A_q(4, 3)$ for general q; and in Chapter 15 we showed that $A_q(n, d) = q^{n-d+1}$ if q is a prime power and $d \leq n \leq q + 1$.

Finally, the problem of finding optimal *linear* codes over $GF(q)$ was considered in Chapters 14 and 15.

A topic not covered in this text is that of *asymptotic bounds*, applicable when n is large. However, much research has been devoted to closing the gap between the best-known asymptotic lower and upper bounds, which are currently an asymptotic version of the Gilbert–Varshamov lower bound (cf. Theorem 8.10) and an upper bound, obtained by linear programming methods, due to McEliece *et al*. (1977). Good accounts of this topic may be found in MacWilliams and Sloane (1977) and van Lint (1982).

We now give brief descriptions of some types of code not previously discussed in this text.

Burst-error correcting codes

The codes we have considered to date are designed to correct *random* errors (e.g. for a binary symmetric channel). It often happens that we need a code for a channel which does not have random errors but which has errors in *bursts*, i.e. several errors close together. There are some linear cyclic codes which are well adapted for burst-error correcting, two important families being *Reed–Solomon codes* and *Fire codes*. An alternative procedure is to scramble the order in which the digits are transmitted, the scrambling occurring over a length of several blocks. Then at the receiving end the order is changed back to the original sequence. This change-back will break up any bursts of errors, leaving errors scattered in a pseudo-random way over several blocks, so that they fall within the capacity of random-error correcting codes. The *interleaving of codes* is one way of carrying out this procedure.

For a good account of burst-error correcting codes, see Peterson and Weldon (1972) or Dornhoff and Hohn (1978).

Convolutional codes

Convolutional codes are powerful error-correcting codes which were introduced by Elias in 1955. They are unlike the codes we have already considered in that message symbols are not broken up into blocks for encoding. Instead check digits are interleaved within a long stream of information digits. For example, for rate $\frac{1}{2}$, one might have the information input $x_1 x_2 x_3 \cdots$ encoded as $x_1 x_1' x_2 x_2' x_3 x_3' \ldots$, where each check digit x_i' is a function of $x_1, x_2, \ldots, x_i$ which is found by means of a feed-back shift register. The decoding is done one digit at a time using the previously received and corrected digits.

Mathematicians tend to be less interested in convolutional codes because the mathematical theory is nothing like as well developed as for block codes. Convolutional codes are also intrinsically more difficult. Despite this, such codes have been extensively used in practice. For example, NASA has been using convolutional codes in deep-space applications since 1977 (from 1969 to 1976, NASA's Mariner-class spacecraft had used a Reed–Muller [32, 6]-block code, as mentioned in Chapter 1).

Chapters on convolutional codes are included in the books by Blahut (1983), McEliece (1977), Peterson and Weldon (1972), and van Lint (1982).

Cryptographic codes

Cryptographic codes have little in common with error-correcting codes, for their aim is the *concealment of information*. The last decade has seen an explosion of interest in such codes following the invention of the concept of the *public-key cipher system* by Diffie and Hellman (1976). Such a system makes use of a *one-way trapdoor function*. This is an encrypting function which has an inverse decrypting function; but if only the encrypting function is known, it is computationally infeasible to discover the decrypting function. This means that a person R can publish his encrypting algorithm (e.g. in a directory) so that *any* member of the public can send messages to R in complete secrecy, for only R knows his own decrypting algorithm. Such a public-key system thus overcomes the weakness of a traditional cipher system which requires the secret delivery of a 'key' in advance of sending a secret message.

Rivest *et al.* (1978) found an elegant way to implement the Diffie–Hellman system by using prime numbers and a simple consequence of Fermat's theorem (Exercise 16.1). Their method relies on the facts that

(a) there are computer algorithms for testing primality which are extremely fast (e.g. a few seconds for a 100-digit number), while

(b) all known algorithms for factorizing composite numbers are extremely slow (e.g. if n is a 200-digit number obtained by multiplying two 100-digit prime numbers, the fastest of today's computers, using the best-known algorithm, would take millions of years to find the prime factors of n).

THE RIVEST–SHAMIR–ADLEMAN (R–S–A) CRYPTOSYSTEM

Let us assume that all messages are encoded as large decimal numbers (e.g. via $A = 01$, $B = 02$, ..., $Z = 26$). The purpose here is not to encrypt the message but merely to get it in the numeric form necessary for encryption.

A subscriber R chooses two large prime numbers p and q, each about 100 digits long, and calculates $n = pq$. He then finds two numbers s and t such that

$$st \equiv 1 \ (\text{mod} \ (p - 1)(q - 1)),$$

i.e. $st = r(p - 1)(q - 1) + 1$, for some integer r.

R publishes the numbers n and s but keeps the numbers p, q, and t secret. He also publishes the encryption algorithm, which is simply:

'encipher a message number x as $y = x^s \ (\text{mod} \ n)$'.

To decipher the received message y, R simply calculates $y^t \ (\text{mod} \ n)$. This gives the original message x because, using Exercise 16.1, we have

$$y^t = x^{st} = x^{r(p-1)(q-1)+1} \equiv x \ (\text{mod} \ n).$$

Remarks (i) A long message number must be broken into blocks, so that each block represents a number smaller than n. The blocks are then enciphered separately.

(ii) Even if n is an enormous number, say 200 digits, a message can be enciphered or deciphered very efficiently, using less than one second of computing time.

(iii) A subscriber R can construct (privately) his key numbers p, q, n, s and t very quickly with a computer. It takes a few seconds to generate a pair of random prime numbers p and q, each having about 100 digits. Then, for a random choice of s, the Euclidean algorithm provides a very fast method of calculating t such that $st \equiv 1 \ (\text{mod} \ (p - 1)(q - 1))$.

(iv) The deciphering procedure is secret because t is known only to R. To find t from n and s requires knowledge of p and q.

This in turn requires factorizing n, which we have already remarked to be computationally infeasible (by known methods).

An illustration of an R–S–A cryptosystem in which p and q are small prime numbers, so that the code may easily be broken, is given in Exercise 16.2.

Interesting expository articles on cryptographic codes are Gardner (1977) and Sloane (1981). For a comprehensive treatment of cipher systems in general, Beker and Piper (1982) is recommended.

Variable-length source codes

In order to illustrate the ideas here, let us consider the problem of transmitting English text over a binary symmetric channel as quickly and as reliably as possible. This can be carried out by applying two codes in series. First a *source code* encodes the text into a long string of binary digits. For reliability, this binary data is then broken into blocks of length k and each block encoded into a codeword of length n by means of an error-correcting $[n, k]$-code. Decoding of the two codes is, of course, done in reverse order.

In choosing the source code we are not concerned with the error-correcting aspects. Our main aim is to encode the source alphabet as economically as possible. If letters in the source alphabet occur with differing frequencies, we can best do this by using a *variable-length source code*.

We now give three examples of source codes for our alphabet of 27 letters ('A' to 'Z' and 'space').

ASCII CODE (AMERICAN STANDARD CODE FOR INFORMATION INTERCHANGE)

Computers are usually constructed internally to handle only 0s and 1s. A source code is therefore required to translate each typed character into a binary vector. A common such code is the ASCII code. This has $128 = 2^7$ codewords representing letters of the alphabet (upper and lower case), digits 0 to 9, and assorted other symbols and instructions. Each codeword is a binary vector of length 7 together with an overall parity check (so that any

Character	Probability	ASCII code	Morse code	Huffman code
space	0.185 9	01000001	space	000
A	0.064 2	10000010	01	0100
B	0.012 7	10000100	1000	0111111
C	0.021 8	10000111	1010	11111
D	0.031 7	10001000	100	01011
E	0.103 1	10001011	0	101
F	0.020 8	10001101	0010	001100
G	0.015 2	10001110	110	011101
H	0.046 7	10010000	0000	1110
I	0.057 5	10010011	00	1000
J	0.000 8	10010101	0111	0111001110
K	0.004 9	10010110	101	01110010
L	0.032 1	10011001	0100	01010
M	0.019 8	10011010	11	001101
N	0.057 4	10011100	10	1001
O	0.063 2	10011111	111	0110
P	0.015 2	10100000	0110	011110
Q	0.000 8	10100011	1101	0111001101
R	0.048 4	10100101	010	1101
S	0.051 4	10100110	000	1100
T	0.079 6	10101001	1	0010
U	0.022 8	10101010	001	11110
V	0.008 3	10101100	0001	0111000
W	0.017 5	10101111	011	001110
X	0.001 3	10110001	1001	0111001100
Y	0.016 4	10110010	1011	001111
Z	0.000 5	10110100	1100	0111001111

Fig. 16.3. Codes for the English alphabet.

single error may be detected). In other words, the ASCII code
is the binary even-weight code of length 8. Those codewords
representing upper case letters are shown in Fig. 16.3.

For other applications, a fixed-length code such as the ASCII
code may be uneconomical.

MORSE CODE

This is a variable-length code which takes advantage of the high
frequency of occurrence of some letters, such as 'E', by making
their codewords short, while very infrequent letters, such as 'Q',

are represented by longer codewords. The Morse code is given in Fig. 16.3, where the 0s may be read as dots and the 1s as dashes. Although the Morse code may appear to be a binary code, it is in fact a ternary code, having the symbols dot, dash, and space. A space has to be left between letters (and at least two spaces between words), for otherwise the code cannot be uniquely decoded; for example, the message 01000110 can mean either LEG or RUN unless spaces are inserted between letters. This drawback means that the Morse code is rarely used nowadays.

HUFFMAN CODES

Suppose a source alphabet has N letters $a_1, a_2, \ldots, a_N$ and that the probability of occurrence of a_i is p_i. Then if each a_i is encoded into a word of length l_i, the average word-length of the code is $\sum_{i=1}^{N} p_i l_i$.

Huffman coding is an ingenious way of matching codewords to source symbols so that

(a)   the code is uniquely decodable, i.e. when any string of source symbols has been encoded into a string of binary digits, it is always clear where one codeword ends and the next one begins, and
(b)   the average word-length is as small as possible.

While omitting the details of how Huffman codes may be constructed, we give an example of such a code for the English alphabet in Fig. 16.3. From the given probabilities, it may be calculated that the average word length is 4.1195. This gives a saving of nearly 18% on the best fixed-length code we could have used, in which all codewords have length 5 (any fixed-length code is clearly uniquely decodable). The reason why a Huffman code is uniquely decodable is that no codeword is a *prefix* of any other codeword, i.e. if $x_1 x_2 \cdots x_n$ is any codeword, then there is no codeword of the form $x_1 x_2 \cdots x_n x_{n+1} \cdots x_m$ for any $m > n$.

For a good account of Huffman source coding, the reader is referred to McEliece (1977), Jones (1979), or Hamming (1980).

Exercises 16

16.1   Suppose p and q are distinct prime numbers. Prove that for any integers x and r,

$$x^{r(p-1)(q-1)+1} \equiv x \pmod{pq}$$

[*Hint:* Use Fermat's theorem: 'if $x \not\equiv 0 \pmod{p}$, then $x^{p-1} \equiv 1 \pmod{p}$' (cf. Exercise 3.8).]

16.2 Suppose a person's published encryption algorithm reads: 'Convert your message to a large decimal number via the code $A = 01$, $B = 02, \ldots, Z = 26$, space $= 00$. Break this number into blocks of length 4. Encipher each block x into the 4-digit block $y = x^{283} \pmod{2813}$'.

Find the decryption algorithm for the above code and hence (with the aid of a pocket calculator) decipher the following intercepted message:

2385 0593 0736 0209 1671 2595 2026 2418.

16.3 In the R–S–A cryptosystem, explain how messages can be 'signed' to prevent forgeries.

16.4 Consider a source alphabet a_1, a_2, a_3, a_4 with probabilities of occurrence $\frac{1}{2}$, $\frac{1}{4}$, $\frac{1}{8}$, $\frac{1}{8}$ respectively. Which of the following source codes are (a) uniquely decodable, (b) prefix-free?

Source letter	p_i	Code A	Code B	Code C	Code D
a_1	0.5	0	00	0	0
a_2	0.25	1	01	10	01
a_3	0.125	00	10	110	011
a_4	0.125	11	11	111	0111

For those codes which are uniquely decodable, calculate the average word-length.

16.5 Use the Huffman code of Fig. 16.3 to decode the message

00101110101000101100101011.

Solutions to exercises

Chapter 1

1.1

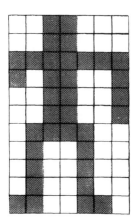

[*Remark*: Pictures have actually been transmitted from Earth into outer space in this way. Two large prime numbers were used so that a much more detailed picture could be sent. It is reasonable to expect that a civilized recipient of such a message would be able to work out how to reconstruct the picture, since factorization of a number into prime factors is a property independent of language or notation.]

1.2 If either 00000 or 11111 is sent, then the received vector will be decoded as the codeword sent if and only if two or fewer errors occur. So the probability that the received vector is corrected to the codeword sent is

$$(1-p)^5 + 5(1-p)^4 p + \binom{5}{2}(1-p)^3 p^2$$
$$= 1 - 10p^3 + 15p^4 - 6p^5,$$

whence the word error probability is $10p^3 - 15p^4 + 6p^5$.

1.3 Suppose $d(C) = 4$. If a received vector **y** has distance ≤ 1 from some codeword, we decode as that codeword. If **y** has distance at least 2 from every codeword, we seek re-transmission. This scheme guarantees the simultaneous correction of single errors and detection of double errors. Note that C could also be used either as a single-error-correcting code or as a *triple*-error-detecting code, but not both *simultaneously* (why not?).

1.4 $\lfloor (16 - 1)/2 \rfloor = 7$.

1.5 Suppose C is a q-ary $(3, M, 2)$-code. Then the M ordered pairs obtained by deleting the third coordinate of each codeword must be distinct (if two such pairs were identical, then the corresponding codewords of C would differ only in the third position, contradicting $d(C) = 2$). So $M \leq q^2$.

A 3-ary $(3, 9, 2)$-code is

$$
\begin{array}{ccc}
0\ 0\ 0 & 1\ 0\ 1 & 2\ 0\ 2 \\
0\ 1\ 1 & 1\ 1\ 2 & 2\ 1\ 0. \\
0\ 2\ 2 & 1\ 2\ 0 & 2\ 2\ 1
\end{array}
$$

More generally it is easily shown that $\{(a, b, a + b) \mid (a, b) \in (F_q)^2\}$, where $F_q = \{0, 1, \ldots, q - 1\}$ and $a + b$ is calculated modulo q, is a q-ary $(3, q^2, 2)$-code.

Chapter 2

2.1 (i) $\{000000, 111111\}$. (ii) $(F_2)^3$. (iii) Add overall parity-check to $(F_2)^3$. (iv) Not possible. Suppose C were a $(5, 3, 4)$-code. There is no loss in assuming 00000 is a codeword. But then the other two codewords each have at least four 1s, which implies that they differ in at most 2 places. (v) Not possible. A binary $(8, M, 3)$-code satisfies the sphere-packing bound, $M(1 + 8) \leq 2^8$, which implies that $M \leq 28$.

2.2 Suppose C is a binary (n, M, d)-code. Partition the codewords of C into two disjoint sets, those ending with a 0 and those ending with a 1. One or other of these sets contains at least $M/2$ of the codewords. Take this set and delete the last coordinate to get an $(n - 1, \geq M/2, d)$-code (this is called a *shortened code* of C). Taking $M = A_2(n, d)$ gives $A_2(n-1, d) \geq \frac{1}{2}A_2(n, d)$.

2.3 Immediate from Exercise 1.5.

2.4 Let C be the code obtained from $(F_2)^{n-1}$ by adding an overall parity check. Every codeword of C has even weight and so $C \subseteq E_n$. Since every vector of E_n may be obtained from one in $(F_2)^{n-1}$ in this way, we have $C = E_n$. Thus $|E_n| = |(F_2)^{n-1}| = 2^{n-1}$. $(F_2)^{n-1}$ has minimum distance 1, and so E_n has minimum distance 2.

2.5 $\binom{50}{8} / \binom{10}{8} > 10^7$.

2.6 Let C be a binary (n, M, d)-code with d even. Delete a suitable coordinate from all codewords to get an $(n-1, M, d-1)$-code and then add an overall parity check (cf. proof of Theorem 2.7).

2.7 Any such code is equivalent to $\{00 \cdots 0, 11 \cdots 100 \cdots 0\}$, where the number of 1s in the second word is one of $1, 2, \ldots, n$.

2.8 Suppose C is a binary $(8, M, 5)$-code, with $M \geq 4$. We may assume $00000000 \in C$. At most one codeword has weight ≥ 6, for two words of weight ≥ 6 could differ in at most four places. So C has at least two codewords of weight 5. Up to equivalence, we may assume these are 11111000 and 11000111. It is now easy to show that the only further codeword possible is 00111111.

2.9 Let C be an (n, q, n)-code over $F_q = \{1, 2, \ldots, q\}$ and let A be a matrix whose rows are the codewords of C. Since $d(C) = n$, the q elements of any column of A must be distinct and so must be precisely the symbols $1, 2, \ldots, q$ in some order. For each column of A a suitable permutation of the symbols may be performed to give

$$
A = \begin{matrix}
1 & 1 & \cdots & 1 \\
2 & 2 & \cdots & 2 \\
\vdots & \vdots & & \vdots \\
q & q & \cdots & q.
\end{matrix}
$$

2.10 Apply either the sphere-packing bound or an argument similar to that of Exercise 1.5 (i.e. the words formed by deleting the last two coordinates must be distinct).

2.11 By Corollary 2.8 and Example 2.23, we have

$$A_2(8, 4) = A_2(7, 3) = 16.$$

2.12 Take as codewords the 11 rows of an incidence matrix of the design, the 11 vectors obtained by interchanging all 0s and 1s, the all-0 vector, and the all-1 vector. The minimum distance may be shown to be 5 by an argument similar to that used in Example 2.23. A binary $(11, M, 5)$-code satisfies

$$M\left[1 + 11 + \binom{11}{2}\right] \leq 2^{11},$$

and so $M \leq 2^{11}/67$, which implies $M \leq 30$.

2.13 (i) Following the hint: for each of the v choices of x there are r choices of B: for each of the b choices of B there are k choices of x. So the number of pairs in the set is $vr = bk$.

 (ii) Let y be a fixed point. Count in two ways the number of ordered pairs in the set

$$\{(x, B): x \text{ is a point}, B \text{ is a block}, x \neq y \text{ and both}$$
$$x \text{ and } y \in B\}.$$

2.14 (i) Condition (ii) of the previous exercise is not satisfied.
 (ii) Immediate from Theorem 2.27(i).

2.15 Easy generalization of the argument of Example 2.23, Exercise 2.12.

2.16 Straightforward check (just 34 comparisons of codewords are required: 11010000 with 19 others, then 11100100 with 11 others, then 10101010 with 3 others, and finally **0** with **1**).

2.17 Since $(\mathbf{u}_1 \mid \mathbf{u}_1 + \mathbf{v}_1) = (\mathbf{u}_2 \mid \mathbf{u}_2 + \mathbf{v}_2)$ if and only if $(\mathbf{u}_1, \mathbf{v}_1) = (\mathbf{u}_2, \mathbf{v}_2)$, the number of codewords in C_3 is $M_1 M_2$.

Let $\mathbf{a} = (\mathbf{u} \mid \mathbf{u} + \mathbf{v})$ and $\mathbf{b} = (\mathbf{u}' \mid \mathbf{u}' + \mathbf{v}')$ be distinct codewords of C_3.

If $\mathbf{v} = \mathbf{v}'$, then $d(\mathbf{a}, \mathbf{b}) = 2d(\mathbf{u}, \mathbf{u}') \geq 2d_1$.

If $\mathbf{v} \neq \mathbf{v}'$, then $d(\mathbf{a}, \mathbf{b}) = d(\mathbf{u}, \mathbf{u}') + d(\mathbf{u} + \mathbf{v}, \mathbf{u}' + \mathbf{v}')$

$$= w(\mathbf{u} + \mathbf{u}') + w(\mathbf{u} + \mathbf{v} + \mathbf{u}' + \mathbf{v}')$$
$$= d(\mathbf{u} + \mathbf{u}', \mathbf{0}) + d(\mathbf{u} + \mathbf{u}', \mathbf{v} + \mathbf{v}')$$
$$\geq d(\mathbf{0}, \mathbf{v} + \mathbf{v}') \quad \text{(by the triangle inequality)}$$
$$= d(\mathbf{v}, \mathbf{v}') \geq d_2.$$

2.18 Let C_1 be the $(8, 128, 2)$-code E_8 (see Exercise 2.4) and let C_2 be the $(8, 20, 3)$-code of Exercise 2.16. Apply Exercise 2.17 to get a $(16, 2560, 3)$-code.

2.19 $C_1 = (4, 8, 2)$-code, $C_2 = (4, 2, 4)$-code $\Rightarrow$
$$C_3 = (8, 16, 4)\text{-code.}$$
$C_1 = (8, 16, 4)$-code, $C_2 = (8, 2, 8)$-code $\Rightarrow$
$$C_3 = (16, 32, 8)\text{-code.}$$
$C_1 = (16, 32, 8)$-code, $C_2 = (16, 2, 16)$-code $\Rightarrow$
$$C_3 = (32, 64, 16)\text{-code.}$$

2.20 Since $w(\mathbf{x}_i + \mathbf{x}_j) = d(\mathbf{x}_i, \mathbf{x}_j) \geq d$, we have

$$w(T) \geq \tfrac{1}{2} M(M-1)d \tag{1}$$

Suppose $\tfrac{1}{2}M - t_j$ codewords have 1 in the jth position, so that $\tfrac{1}{2}M + t_j$ codewords have 0 in the jth position. Then the number of 1s in the jth column of T is

$$(\tfrac{1}{2}M - t_j)(\tfrac{1}{2}M + t_j) = (\tfrac{1}{2}M)^2 - t_j^2$$
$$\leq \begin{cases} (\tfrac{1}{2}M)^2 & \text{if } M \text{ is even} \\ (\tfrac{1}{2}M)^2 - \tfrac{1}{4} & \text{if } M \text{ is odd,} \end{cases}$$

since $t_j^2 \geq (\tfrac{1}{2})^2$ if M is odd. Hence

$$w(T) \leq \begin{cases} \tfrac{1}{4}M^2 n & \text{if } M \text{ is even} \\ \tfrac{1}{4}(M^2 - 1)n & \text{if } M \text{ is odd} \end{cases} \tag{2}$$

(1) and (2) give the required result.

2.21 If $A_2(n, d)$ is even, the result is immediate. If $A_2(n, d)$ is odd, use $\lfloor 2x \rfloor \leq 2\lfloor x \rfloor + 1$.
     The result gives $A_2(9, 5) \leq 10$ and $A_2(10, 6) \leq 6$. The former bound can be improved via Corollary 2.8 and the latter bound; thus $A_2(9, 5) = A_2(10, 6) \leq 6$.

2.22 (i) was shown in Exercise 2.21. (ii) follows from (i) and Corollary 2.8. (iii) By (i), $A_2(2d - 1, d) \leq 2d$. Hence $A_2(2d, d) \leq 4d$ by Exercise 2.2. (iv) follows from (iii) and Corollary 2.8.

2.23 The $(32, 64, 16)$-code is optimal by Exercise 2.22 (iii). The generalization follows from the Remark in Exercise 2.19 and Exercise 2.22 (iii).

2.24 Immediate from Exercises 2.15 and 2.22(iii).

Chapter 3

3.1 $2^{20} \equiv (2^3)^6 2^2 \equiv 1^6 2^2 \equiv 4 \pmod 7$.
$3^{100} \equiv (3^4)^{25} \equiv 1^{25} \equiv 1 \pmod{10}$.

3.2 $x \equiv 0, 1, 2$ or $3 \pmod 4 \Rightarrow x^2 \equiv 0, 1, 0$ or $1 \pmod 4$ respectively. Hence $x^2 + y^2 \equiv 0, 1$ or $2 \pmod 4$, but $1839 \equiv 3 \pmod 4$.

3.3 x: 1 2 3 4 5 6   1 2 3 4 5 6 7 8 9 10 11 12
x^{-1}: 1 4 5 2 3 6   1 7 9 10 8 11 2 5 3 4 6 12

3.4 (i) 2, (ii) $\frac{1}{11}$.

3.5 Yes, No, No.

3.6 (i) $1 \cdot 0 + 2 \cdot 1 + 3 \cdot 3 + 4 \cdot 1 + 5x + 6 \cdot 9 + 7 \cdot 1 + 8 \cdot 3 + 9 \cdot 9 + 10 \cdot 9 \equiv 0 \pmod{11} \Rightarrow 5x + 7 \equiv 0 \Rightarrow 5x \equiv 4 \Rightarrow x \equiv 4 \cdot 5^{-1} \equiv 4 \cdot 9 \equiv 3$.

(ii) The number is 00232xy800, where we see that each of x, y is 0, 8 or 9. For the number to be an ISBN, we require $6x + 7y \equiv 7$, i.e. $y = 1 + 7x$. Now $x = 0 \Rightarrow y = 1$; $x = 8 \Rightarrow y = 2$; $x = 9 \Rightarrow y = 9$. So $x = y = 9$.

3.7 Suppose $x_1 \cdots x_{10}$ is the codeword sent and $y_1 \cdots y_{10}$ the vector received. If a single error has occurred of magnitude a, then $\sum_{i=1}^{10} y_i = (\sum_{i=1}^{10} x_i) + a \equiv a \pmod{11}$. So the error is detected. Unlike the ISBN code, any transposition of two digits will go undetected, for then $\sum y_i = \sum x_i \equiv 0$.

3.8 $1a, 2a, \ldots, (p-1)a$ are distinct $\pmod p$, for $ia \equiv ja \pmod p \Rightarrow i \equiv j \pmod p$ (multiplying both sides by a^{-1}). So $1a, 2a, \ldots, (p-1)a$ are congruent to the elements $1, 2, \ldots, p-1$ in some order. Hence $1a \cdot 2a \cdots \cdot (p-1)a \equiv 1 \cdot 2 \cdots \cdot (p-1) \pmod p$ and so $(p-1)! \, a^{p-1} \equiv (p-1)! \pmod p$. Multiplying through by the inverse of $(p-1)!$ gives $a^{p-1} \equiv 1 \pmod p$.

3.9 $a \not\equiv 0 \Rightarrow \gcd(a, p) = 1$. By the Euclidean algorithm, $1 = ax + py$ for some integers x and y. Hence $ax \equiv 1 \pmod p$ and so $x = a^{-1}$. $31 = 1 \cdot 23 + 8$; $23 = 2 \cdot 8 + 7$; $8 = 1 \cdot 7 + 1$. So $1 = 8 - 1 \cdot 7 = 3 \cdot 8 - 23 = 3 \cdot 31 - 4 \cdot 23$. Hence $-4 \cdot 23 \equiv 1 \pmod{31}$ and so $23^{-1} = -4 \equiv 27$.

3.10 2, 3, 2 (other answers possible).

3.11 Let 1 be the multiplicative identity element of F. The field elements $n1$ for $n = 1, 2, 3, \ldots$ cannot all be distinct, since F is finite. So $l1 = m1$ for some $0 < m < l$, whence $(l - m)1 = 0$. This implies that $n1 = 0$ for some integer n. Let p

be the smallest positive integer such that $p1 = 0$. Then p is prime because $p = rs$, with $1 < r, s < p$, $\Rightarrow p1 = (r1)(s1) = 0 \Rightarrow r1 = 0$ or $s1 = 0$ (by Lemma 3.1 (ii)), contradicting the minimality of p. Finally, if $\alpha \in F$, then $p\alpha = \alpha + \alpha + \cdots + \alpha = \alpha(1 + 1 + \cdots + 1) = \alpha(p1) = \alpha 0 = 0$.

3.12 $\binom{p}{i} = p!/i! \, (p - i)!$. If $i \in \{1, 2, \ldots, p - 1\}$, then the numerator $p!$ is divisible by p, whereas the denominator $i! \, (p - i)!$ is not. Hence $\binom{p}{i} \equiv 0 \, (\mathrm{mod} \, p)$. By the binomial theorem,

$$(a + b)^p \equiv \sum_{i=0}^{p} \binom{p}{i} a^i b^{p-i} \equiv a^p + b^p \, (\mathrm{mod} \, p)$$

For the last part, use induction on a.

3.13   In the product, each element x will cancel with its inverse, except when $x = x^{-1}$. Now $x = x^{-1} \Leftrightarrow x^2 = 1 \Leftrightarrow (x - 1)(x + 1) = 0 \Leftrightarrow x = 1$ or $x = -1$.

3.14   (i) $x^2 = y^2 \Rightarrow (x - y)^2 = x^2 - y^2 = 0 \Rightarrow x - y = 0 \Rightarrow x = y$.

So the squares of the non-zero elements are precisely the distinct non-zero elements (in some order).

(ii)   *Hint:* show that if $a \neq 0$, then $x^2 = a$ has either 2 or 0 solutions.

Chapter 4

4.1   Show that this single condition holds if and only if both conditions (1) and (2) of Theorem 4.1 hold.

4.2   Suppose $\mathbf{x}, \mathbf{y} \in E_n$, so that $w(\mathbf{x})$ and $w(\mathbf{y})$ are even numbers. By Lemma 2.6, $w(\mathbf{x} + \mathbf{y}) = w(\mathbf{x}) + w(\mathbf{y}) - 2w(\mathbf{x} \cap \mathbf{y}) =$ an even number. So $\mathbf{x} + \mathbf{y} \in E_n$ and hence E_n is a subspace. By Exercise 2.4, $|E_n| = 2^{n-1}$ and so $\dim (E_n) = n - 1$.

The rows of

$$\begin{array}{l} 1\ 0\ 0\ \cdots\ 0\ 1 \\ 0\ 1\ 0\ \cdots\ 0\ 1 \\ \vdots \\ 0\ 0\ \cdots\ 0\ 1\ 1 \end{array}$$

form a basis (other answers are possible).

4.3 $(1, 2, 0, 1) = 2(0, 1, 2, 1) + (1, 0, 2, 2)$.   So $\{(0, 1, 2, 1),$
       $(1, 0, 2, 2)\}$ is a basis and dim $(C) = 2$.

4.4   Show that $\{\mathbf{u}, \mathbf{v}\}$ is linearly dependent if and only if either
       $\mathbf{u}$ or $\mathbf{v}$ is zero or $\mathbf{v}$ is a scalar multiple of $\mathbf{u}$.

4.5   In each case show that the new set is still both a spanning
       set and a linearly independent set.

4.6   In F, let n denote the element $1 + 1 + \cdots + 1$ (n1s). Then
       the subset $\{0, 1, \ldots, p - 1\}$ of F may be regarded as the
       field $GF(p)$, since addition and multiplication are carried
       out modulo p. It follows at once that F is a vector space
       over $GF(p)$, all the axioms following immediately from
       the field properties of F and $GF(p)$. If the vector space F
       over $GF(p)$ has dimension h, then it follows, as in the
       proof of Theorem 4.3, that $|F| = p^h$.

4.7   We omit the proof of the general result here, as it will be
       given in Chapter 14. The points of P_2 are $\{000, 100\}$,
       $\{000, 010\}$, $\{000, 001\}$, $\{000, 110\}$, $\{000, 101\}$, $\{000, 011\}$,
       and $\{000, 111\}$. The lines are $\{000, 100, 010, 110\}$,
       $\{000, 100, 001, 101\}$, etc. That this 7-point plane is the
       same as that of Example 2.19 may be seen from Fig. 14.8,
       wherein a vector $\mathbf{x}$ stands for the point $\{0, \mathbf{x}\}$.

Chapter 5

5.1   No; 24 is not a power of 2.

5.2 $[n, n - 1, 2]$, $\begin{bmatrix} & & 1 \\ & & 1 \\ I_{n-1} & & \vdots \\ & & \vdots \\ & & 1 \end{bmatrix}$.

5.3   We use Theorem 4.1.

$$\mathbf{x}, \mathbf{y} \in C \Rightarrow (\mathbf{x} + \mathbf{y})H^T = \mathbf{x}H^T + \mathbf{y}H^T = 0 + 0 = 0$$
$$\Rightarrow \mathbf{x} + \mathbf{y} \in C.$$
$$\mathbf{x} \in C \text{ and } a \in GF(q) \Rightarrow (a\mathbf{x})H^T = a(\mathbf{x}H^T) = a0 = 0$$
$$\Rightarrow a\mathbf{x} \in C.$$

5.4   If $\mathbf{x} = (x_1, \ldots, x_n) \in C$, let $\hat{\mathbf{x}} = (x_1, \ldots, x_n, \sum_{i=1}^{n} x_i)$, where
       $\sum x_i$ is calculated modulo 2. Then $\hat{C} = \{\hat{\mathbf{x}} \mid \mathbf{x} \in C\}$. Suppose

C is linear, so that $\mathbf{x}, \mathbf{y} \in C \Rightarrow \mathbf{x} + \mathbf{y} \in C$. Then

$$\hat{\mathbf{x}}, \hat{\mathbf{y}} \in \hat{C} \Rightarrow \hat{\mathbf{x}} + \hat{\mathbf{y}} = \left(x_1 + y_1, \ldots, x_n + y_n, \sum x_i + \sum y_i \right)$$

$$= \left(x_1 + y_1, \ldots, x_n + y_n, \sum (x_i + y_i) \right)$$

$$= (\widehat{\mathbf{x} + \mathbf{y}}) \in \hat{C}.$$

So $\hat{C}$ is linear.

Adding an overall parity check to the code of Example 5.6(ii) gives an $[8, 4, 4]$-code with generator matrix

$$\begin{bmatrix} 1 & 0 & 0 & 0 & 1 & 0 & 1 & 1 \\ 0 & 1 & 0 & 0 & 1 & 1 & 1 & 0 \\ 0 & 0 & 1 & 0 & 1 & 1 & 0 & 1 \\ 0 & 0 & 0 & 1 & 0 & 1 & 1 & 1 \end{bmatrix}.$$

5.5 Let Ev and Od denote the subsets of C consisting of words of even and odd weights respectively. Suppose $Ev \neq C$. Then there exists a codeword, $\mathbf{y}$ say, of odd weight. Now the set $Ev + \mathbf{y} = \{\mathbf{x} + \mathbf{y} \mid \mathbf{x} \in Ev\}$ is contained in C (since C is linear). But all words in $Ev + \mathbf{y}$ are odd (via $w(\mathbf{x} + \mathbf{y}) = w(\mathbf{x}) + w(\mathbf{y}) - 2w(\mathbf{x} \cap \mathbf{y})$, cf. Lemma 2.6), and so we have $Ev + \mathbf{y} \subseteq Od$. Hence $|Ev| = |Ev + \mathbf{y}| \leq |Od|$. Also $Od + \mathbf{y} \subseteq Ev$ and so $|Od| \leq |Ev|$. Hence $|Ev| = |Od| = \frac{1}{2}|C|$.

5.6

$$C_1 = \begin{matrix} 0 & 0 & 0 & 0 & 0 \\ 1 & 1 & 1 & 1 & 0 \\ 0 & 0 & 1 & 1 & 1 \\ 1 & 1 & 0 & 0 & 1 \end{matrix}$$

$d(C_1) = $ minimum non-zero weight $= 3$

$$C_2 = \begin{matrix} 0 & 0 & 0 & 0 & 0 & 0 & 0 \\ 1 & 0 & 0 & 1 & 1 & 0 & 1 = \mathbf{x}_1 \\ 0 & 1 & 0 & 1 & 0 & 1 & 1 = \mathbf{x}_2 \\ 0 & 0 & 1 & 0 & 1 & 1 & 1 = \mathbf{x}_3 \\ 1 & 1 & 0 & 0 & 1 & 1 & 0 = \mathbf{x}_1 + \mathbf{x}_2 \\ 1 & 0 & 1 & 1 & 0 & 1 & 0 = \mathbf{x}_1 + \mathbf{x}_3 \\ 0 & 1 & 1 & 1 & 1 & 0 & 0 = \mathbf{x}_2 + \mathbf{x}_3 \\ 1 & 1 & 1 & 0 & 0 & 0 & 1 = \mathbf{x}_1 + \mathbf{x}_2 + \mathbf{x}_3 \end{matrix}$$

$d(C_2) = 4$.

5.7   0 0 0 0 $d(C)$ = minimum non-zero
      1 0 1 1 = $\mathbf{x}_1$ weight = 3.
                                        Since $9[1 + 2 \cdot 4] = 3^4$, the sphere-
      0 1 1 2 = $\mathbf{x}_2$ packing bound is attained and so C
      2 0 2 2 = $2\mathbf{x}_1$ is perfect.
      0 2 2 1 = $2\mathbf{x}_2$
      1 1 2 0 = $\mathbf{x}_1 + \mathbf{x}_2$
      2 2 1 0 = $2\mathbf{x}_1 + 2\mathbf{x}_2$
      1 2 0 2 = $\mathbf{x}_1 + 2\mathbf{x}_2$
      2 1 0 1 = $2\mathbf{x}_1 + \mathbf{x}_2$

5.8   By Table 2.4, $A_2(8, 3) = 20$, $A_2(8, 4) = 16$, and $A_2(8, 5) =$
      4. By Exercise 5.4(ii), there exists a linear [8, 4, 4]-code
      and so $B_2(8, 4) = 16$. There certainly exists also an [8, 4, 3]-
      code and so, since $B_2(8, 3)$ is a power of 2 and is $\leqslant 20$, we
      have $B_2(8, 3) = 16$. The code constructed in Exercise 2.8 is
      linear and so $B_2(8, 5) = 4$.

5.9 $\begin{bmatrix} 1 & 0 & 1 \\ 0 & 1 & 1 \end{bmatrix}$ generates a [3, 2, 2]-code over $GF(q)$.

5.10 First get the required permutation of the rows of A by
      permuting the rows of G. The I_k part will have been
      disturbed but can be restored by a suitable permutation of
      the first k columns.

5.11 $\begin{bmatrix} 1 & 0 & 0 & 0 & 0 & 1 & 1 \\ 0 & 1 & 0 & 0 & 1 & 0 & 1 \\ 0 & 0 & 1 & 0 & 1 & 1 & 0 \\ 0 & 0 & 0 & 1 & 1 & 1 & 1 \end{bmatrix}$.
      No, Yes (by Exercise 5.10).

5.12 $(\mathbf{u} \mid \mathbf{u} + \mathbf{v}) + (\mathbf{u}' \mid \mathbf{u}' + \mathbf{v}') = (\mathbf{u} + \mathbf{u}' \mid \mathbf{u} + \mathbf{u}' + \mathbf{v} + \mathbf{v}') \in C_3$.
      Thus C_3 is linear. So $B_2(2d, d) = 4d$ by Exercises 2.19 and
      2.23 since, at each step, C_1 and C_2 are linear.

Chapter 6

6.1 C_1: $\boxed{00 \ 01 \ 10 \ 11}$ C_2: $\boxed{\begin{array}{l} 000 \ 101 \ 011 \ 110 \\ 100 \ 001 \ 111 \ 010 \end{array}}$

C_3:

$$\begin{array}{llll}
00000 & 10110 & 01011 & 11101 \\
10000 & 00110 & 11011 & 01101 \\
01000 & 11110 & 00011 & 10101 \\
00100 & 10010 & 01111 & 11001 \\
00010 & 10100 & 01001 & 11111 \\
00001 & 10111 & 01010 & 11100 \\
11000 & 01110 & 10011 & 00101 \\
10001 & 00111 & 11010 & 01100
\end{array}$$

(i) 11101, 01011
(ii) e.g. (a) 00000 received as 11000, (b) 00000 received as 10100.

6.2 $P_{corr}(C_1) = (1-p)^2 = 0.9801$
$P_{corr}(C_2) = (1-p)^3 + p(1-p)^2 = (1-p)^2 = 0.9801$
$P_{corr}(C_3) = (1-p)^3(1-2p^2+3p) \approx 0.9992$
There is no point in using C_2 for error correction since P_{corr} is the same as for C_1, while C_2 takes 50% longer than C_1 to transmit messages. C_3 reduces the word error rate considerably.
$P_{undetec}(C_1) = 2p(1-p) + p^2 = 0.0199$
$P_{undetec}(C_2) = 3p^2(1-p) \approx 0.000297$
$P_{undetec}(C_3) = 2p^3(1-p)^2 + (1-p)p^4 \approx 0.00000197.$

6.3 (i) No, communication is impossible.
(ii) Yes, interchange all 0s and 1s in the received vector before decoding.

6.4 The coset leaders include all vectors of weight $\leq t$ and α_{t+1} vectors of weight $t+1$. So the probability that the error vector is *not* a coset leader is

$$\left[\binom{n}{t+1} - \alpha_{t+1}\right]p^{t+1}(1-p)^{n-t-1} + \text{terms involving } p^{t+2}$$

and higher powers. Hence

$$P_{err} \simeq \left[\binom{n}{t+1} - \alpha_{t+1}\right]p^{t+1} \text{ for small } p.$$

6.5 Straightforward calculation, with $A_3 = A_4 = 7$, $A_7 = 1$.

6.6   Since the code is perfect 3-error-correcting, we have

$$\alpha_0 = 1, \qquad \alpha_1 = 23, \qquad \alpha_2 = \binom{23}{2}, \qquad \alpha_3 = \binom{23}{3},$$

and

$$\alpha_i = 0 \text{ for } i \geqslant 4.$$

$$P_{\text{corr}} = (1-p)^{20}(1540p^3 + 210p^2 + 20p + 1) \simeq 0.99992$$

if $p = 0.01$.
So $P_{\text{err}} \simeq 0.000\,08$ [*Remark:* A fair approximation is obtained by using Exercise 6.4; namely $\binom{23}{4}10^{-8}$.]

6.7   Suppose $\mathbf{x} = x_1 x_2 \cdots x_n$ is sent and that the received vector is decoded as $\mathbf{x}' = x_1' x_2' \cdots x_n'$. Then

$$P_{\text{symb}} = \frac{1}{k}\sum_{j=1}^{k} \text{Prob}\,(x_j' \neq x_j)$$

$$= \frac{1}{k}\sum_{\mathbf{e} \in V(n,2)} f(\mathbf{e})\,\text{Prob}\,(\mathbf{e} \text{ is error vector}),$$

where $f(\mathbf{e}) = $ number of incorrect information symbols after decoding if the error vector is $\mathbf{e}$, and so

$$P_{\text{symb}} = \frac{1}{k}\sum_{i=1}^{2^k} F_i P_i.$$

6.8 $$P_{\text{symb}} = \tfrac{1}{2}[P_2 + P_3 + 2P_4]$$
$$= \tfrac{1}{2}[\{2(1-p)^2 p^2 + (1-p)p^3 + p^4\}$$
$$+ \{(1-p)^3 p + (1-p)^2 p^2 + 2(1-p)p^3\}$$
$$+ 2\{3(1-p)^2 p^2 + (1-p)p^3\}].$$

6.9   Note that $P_{\text{err}} = \sum_{i=2}^{2^k} P_i$. Since $F_1 = 0$ and $1 \leqslant F_i \leqslant k$ for all $i \geqslant 2$, we have

$$\frac{1}{k}\sum_{i=2}^{2^k} P_i \leqslant \frac{1}{k}\sum_{i=1}^{2^k} F_i P_i \leqslant \sum_{i=2}^{2^k} P_i$$

and hence the result.

Chapter 7

7.1
$$\mathbf{u} \cdot \mathbf{v} = \sum_{i=1}^{n} u_i v_i = \sum_{i=1}^{n} v_i u_i = \mathbf{v} \cdot \mathbf{u}.$$

$$(\lambda \mathbf{u} + \mu \mathbf{v}) \cdot \mathbf{w} = \sum_{i=1}^{n} (\lambda u_i + \mu v_i) w_i = \sum_{i=1}^{n} (\lambda u_i w_i + \mu v_i w_i)$$
$$= \lambda \sum_{i=1}^{n} u_i w_i + \mu \sum_{i=1}^{n} v_i w_i = \lambda \mathbf{u} \cdot \mathbf{w} + \mu \mathbf{v} \cdot \mathbf{w}.$$

7.2 The standard form generator matrix of E_n was found in Exercise 5.2. It follows from this and Theorem 7.6 that a generator matrix for $E_n^\perp$ is $[11 \cdots 1]$. So $E_n^\perp = \{00 \cdots 0, 11 \cdots 1\}$, which is the repetition code of length n.

7.3 Find the syndrome $S(\mathbf{y})$ of the received vector $\mathbf{y}$. If $S(\mathbf{y}) = \mathbf{0}$, then $\mathbf{y}$ is a codeword. If $S(\mathbf{y}) \neq \mathbf{0}$, then $\mathbf{y}$ is not a codeword and we have detected errors.

7.4 Suppose $\mathbf{x}$ is the codeword sent and $\mathbf{y} = \mathbf{x} + \mathbf{e}$ is received, where $\mathbf{e} = e_1 e_2 \cdots e_n$ is the error vector. Then $S(\mathbf{y}) = (\mathbf{x} + \mathbf{e}) H^T = \mathbf{x} H^T + \mathbf{e} H^T = \mathbf{e} H^T$. So $S(\mathbf{y})^T = H \mathbf{e}^T = \sum_{j=1}^{n} e_j \mathbf{H}_j$, where $\mathbf{H}_j$ is the jth column of H.

7.5 Since the code is perfect, the coset leaders are precisely those vectors of weight $\leqslant 1$. G is in the form $[I_4 \mid A]$ and so

$$H = [-A^T \mid I_3] = \begin{bmatrix} 1110100 \\ 1101010 \\ 1011001 \end{bmatrix}.$$

We use this to construct the syndrome look-up table:

Syndrome	coset leader
000	0000000
111	1000000
110	0100000
101	0010000
011	0001000
100	0000100
010	0000010
001	0000001

$S(0000011) = 011$; decode as

$$0000011 - 0001000 = 0001011.$$

The other three vectors are decoded as 1111111, 0100110, 0010101.

7.6 (a) $\begin{bmatrix} 1022 \\ 0121 \end{bmatrix}$ (b) $\begin{bmatrix} 1110 \\ 1201 \end{bmatrix}$.

(c) A listing of the codewords reveals that $d(C) = 3$. So the 9 vectors of weight $\leqslant 1$ are all coset leaders. Since the total number of coset leaders $= 3^4/3^2 = 9$, the vectors of weight $\leqslant 1$ are *precisely* the coset leaders (in fact the code is perfect). The look-up table is now easily constructed, and the given vectors decoded as

0121, 1201, 2220.

7.7 0612960587.

7.8 Let C be a q-ary $(10, M, 3)$-code. Consider the M vectors of length 8 obtained by deleting the last two coordinates. These vectors must be distinct (or the corresponding vectors of C would be distance $\leqslant 2$ apart). So $M \leqslant q^8$ (this is a particular case of the Singleton bound, Theorem 10.17). In particular, $A_{10}(10, 3) \leqslant 10^8$, $A_{11}(10, 3) \leqslant 11^8$. [*Remark:* The sphere-packing bound is not as good in these cases.] We have $A_{11}(10, 3) = 11^8$ because the linear $[10, 8]$-code over $GF(11)$ having

$$H = \begin{bmatrix} 111 \cdots 1 \\ 123 \cdots 10 \end{bmatrix}$$

is an 11-ary $(10, 11^8, 3)$-code.

7.9 For example, **0** and 0505000000 are codewords only distance 2 apart.

7.10 Let $\mathbf{e}_j = 0 \cdots 01 \cdots 1$ (j1s). We require a code such that $\mathbf{e}_0, \mathbf{e}_1, \ldots, \mathbf{e}_7$ are all in different cosets (we could then decode via syndrome decoding with the $\mathbf{e}_i$s as coset leaders). This requires that $2^7/2^k \geqslant 8$, i.e. $k \leqslant 4$, and so the rate cannot be greater than $\frac{4}{7}$. To achieve rate $\frac{4}{7}$ we would need a 3×7 parity-check matrix H such that $\mathbf{e}_i H^T \neq \mathbf{e}_j H^T$ if $i \neq j$, i.e. such that $(\mathbf{e}_i - \mathbf{e}_j)H^T \neq \mathbf{0}$ for all $i \neq j$. Note that each $\mathbf{e}_i - \mathbf{e}_j$ is a vector of the form $0 \cdots 01 \cdots 10 \cdots 0$. A suitable H is

$$\begin{bmatrix} 0001000 \\ 0100010 \\ 1010101 \end{bmatrix}.$$

If $\mathbf{e}_i - \mathbf{e}_j$ is orthogonal to the first row of H, then all its 1s are to the left or to the right of centre. If also $\mathbf{e}_i - \mathbf{e}_j$ is orthogonal to the second row of H, then there can only be one 1, in one of the 1st, 3rd, 5th or 7th positions. But then $\mathbf{e}_i - \mathbf{e}_j$ is not orthogonal to the third row of H. (*Note:* a similar code of maximum possible rate may be constructed of *any* given length.)

7.11 If C is an $[n, k]$-code, then $\hat{C}$ is an $[n + 1, k]$-code and so a parity-check matrix of $\hat{C}$ is an $(n + 1 - k) \times (n + 1)$ matrix whose rows form a linearly independent set of codewords in $\hat{C}^{\perp}$. It is easily seen that $\bar{H}$ is such a matrix.

Chapter 8

8.1 $H = \begin{bmatrix} 000000011111111 \\ 000111100001111 \\ 011001100110011 \\ 101010101010101 \end{bmatrix}.$

When $\mathbf{y}$ is received, calculate $\mathbf{y}H^T$; this gives the binary representation of the assumed error position. If two or more errors have occurred, then $\mathbf{y}$ will be decoded as a codeword different from that sent.

8.2 11100001, 01111000, at least two errors, 00110011.

8.3 From the standard form generator matrix (see Example 5.6(ii)), write down a parity-check matrix (via Theorem 7.6) and observe that its columns are the non-zero vectors of $V(3, 2)$.

8.4 For C, $\alpha_0 = 1$ and $\alpha_1 = n$, giving $P_{\text{corr}}(C) = (1 - p)^{n-1}(1 - p + np)$. Because every vector in $V(n, 2)$ has distance $\leqslant 1$ from a codeword of C, it follows that every vector in $V(n + 1, 2)$ has distance $\leqslant 2$ from a codeword of $\hat{C}$. Consequently, the coset leaders for $\hat{C}$ all have weight $\leqslant 2$

and so $\alpha_0 = 1$, $\alpha_1 = n + 1$, $\alpha_2 = n$, which leads to $P_{\text{corr}}(\hat{C}) = (1 - p)^{n-1}(1 - p + np)$. [*Remark:* This result will be generalized to any perfect binary code in Exercise 9.1.]

8.5 (i) $\begin{bmatrix} 0 & 1 & 1 & 1 & 1 & 1 & 1 \\ 1 & 0 & 1 & 2 & 3 & 4 & 5 & 6 \end{bmatrix}$, 35234106, 10561360.

(ii) $\begin{bmatrix} 0 & 0 & 0 & 0 & 0 & 0 & 1 & 1 & 1 & 1 & 1 & 1 & 1 & 1 \\ 0 & 1 & 1 & 1 & 1 & 1 & 0 & 0 & 0 & 0 & 0 & 1 & 1 & 1 & 1 \\ 1 & 0 & 1 & 2 & 3 & 4 & 0 & 1 & 2 & 3 & 4 & 0 & 1 & 2 & 3 \end{bmatrix}$

$\begin{bmatrix} 1 & 1 & 1 & 1 & 1 & 1 & 1 & 1 & 1 & 1 & 1 & 1 & 1 & 1 & 1 \\ 1 & 2 & 2 & 2 & 2 & 2 & 3 & 3 & 3 & 3 & 4 & 4 & 4 & 4 & 4 \\ 4 & 0 & 1 & 2 & 3 & 4 & 0 & 1 & 2 & 3 & 4 & 0 & 1 & 2 & 3 & 4 \end{bmatrix}$

8.6 3.

8.7 $\begin{bmatrix} 0 & 4 & 3 & 1 & 0 \\ 3 & 2 & 0 & 0 & 1 \end{bmatrix}$, $\begin{bmatrix} 1 & 1 & 1 & 0 & 0 & 0 \\ 1 & 0 & 0 & 2 & 1 & 0 \\ 0 & 2 & 0 & 1 & 0 & 1 \end{bmatrix}$

[For the code C_2, a column operation (e.g. interchange of columns 3 and 4) is necessary during the reduction of G to a standard form of G'. So, after applying Theorem 7.6 to get a parity check matrix H' corresponding to G', the above column operation must be reversed in H' in order to get a parity-check matrix for the original code C_2.]

$d(C_1) = 2$, $d(C_2) = 3$.

(other answers possible)

8.8 For example,

$$H = \begin{bmatrix} 1 & 0 & 0 & 1 & 1 & 1 \\ 0 & 1 & 0 & 1 & 2 & 3 \\ 0 & 0 & 1 & 1 & 3 & 4 \end{bmatrix}$$

has the property that any three columns form a linearly independent subset of $V(3, 5)$, and so H is the parity-check matrix of a $[6, 3, 4]$-code.

8.9 $R_r = \dfrac{k}{n} = (2^r - 1 - r)/(2^r - 1) = 1 - \dfrac{r}{2^r - 1} \rightarrow 1$ as $r \rightarrow \infty$.

8.10 As in solution to Exercise 7.8, $A_q(n, 3) \leqslant q^{n-2}$. Now suppose q is a prime power. Then the bound is achieved for $n = q + 1$ by Ham $(2, q)$ and for $n < q + 1$ by shortenings of Ham $(2, q)$.

8.11 $f(t) =$ least value of M for which there exists a ternary code of length t with M codewords such that any vector in $V(t, 3)$ has distance $\leqslant 1$ from at least one codeword. For such a code the spheres of radius 1 about codewords must 'cover' the whole space $V(t, 3)$ and so a lower bound on M is given by

$$M(1 + 2t) \geqslant 3^t \tag{1}$$

(This is the sphere-packing bound, but with the inequality reversed.)

(a) (i) If $t = (3^r - 1)/2$ then (1) gives $f(t) \geqslant 3^{t-r}$. The bound is achieved by a perfect $[t, t-r, 3]$-Hamming code over $GF(3)$. So, for $t = (3^r - 1)/2$, we have $f(t) = 3^{t-r}$.

(ii) Generating Ham $(2, 3)$ by $\begin{bmatrix} 1011 \\ 0112 \end{bmatrix}$ and replacing '0' by 'X', we get the entry

X 1 X 2 X 1 2 1 2
X X 1 X 2 1 2 2 1
X 1 1 2 2 2 1 X X
X 1 2 2 1 X X 2 1

(b) The lower bound $f(5) \geqslant 23$ is given by (1). A crude upper bound is $f(5) \leqslant 27$. This is obtained by combining each of the 9 bets for $t = 4$ with each of the forecasts 1, 2, X for the 5th match. The surprising result proved by Kamps and van Lint is that one cannot do better than this.

8.12 Let C be an (n, M, d)-code with $M = A_q(n, d)$. Then there is no vector in $V(n, q)$ with distance $\geqslant d$ from all codewords in C. Thus the spheres of radius $d - 1$ about codewords cover $V(n, q)$, whence the result. (The proof shows that a code meeting the lower bound may be constructed simply by starting with any word and then successively adding new words which have distance at least d from the words already chosen).

Chapter 9

9.1 Suppose C is a perfect t-error-correcting $[n, k]$-code, so that

$$\sum_{i=0}^{t} \binom{n}{i} = 2^{n-k}.$$

As in Exercise 8.4, for $\hat{C}$,

$$\alpha_i = \binom{n+1}{i} \text{ for } 0 \leqslant i \leqslant t,$$

and

$$\alpha_{t+1} = 2^{n+1-k} - \sum_{i=0}^{t} \binom{n+1}{i}$$

$$= 2 \cdot \sum_{i=0}^{t} \binom{n}{i} - \sum_{i=0}^{t} \binom{n}{i} - \sum_{i=1}^{t} \binom{n}{i-1} = \binom{n}{t}.$$

Hence

$$P_{\text{corr}}(\hat{C}) = \sum_{i=0}^{t} \binom{n+1}{i} p^i (1-p)^{n+1-i} + \binom{n}{t} p^{t+1} (1-p)^{n-t}$$

$$= \sum_{i=0}^{t} \binom{n}{i} p^i (1-p)^{n+1-i}$$

$$+ \sum_{i=1}^{t} \binom{n}{i-1} p^i (1-p)^{n+1-i} + \binom{n}{t} p^{t+1} (1-p)^{n-t}$$

$$= (1-p) P_{\text{corr}}(C)$$

$$+ \left(p\, P_{\text{corr}}(C) - \binom{n}{t} p^{t+1} (1-p)^{n-t} \right)$$

$$+ \binom{n}{t} p^{t+1} (1-p)^{n-t}$$

$$= P_{\text{corr}}(C).$$

9.2 It is easily checked that $\mathbf{u} \cdot \mathbf{v} = 0$ for any rows $\mathbf{u}$ and $\mathbf{v}$ of G. It follows that $G_{12}^{\perp} = G_{12}$. Now show that G_{12} has no codeword of weight $\leqslant 5$ by imitating the proof of Lemma 3 in the proof of Theorem 9.3.

9.3 If $H = [I_5 \mid A]$ has no 4 columns linearly dependent, then each column of A has at most one zero, and no two columns of A can have a zero entry in common (or their sum or difference would be a linear combination of two of the columns of I_5). The hint now follows easily. It then follows that in each of the undecided columns of A, two of the *s are 2s and the other * is a 1. The remaining columns may now be completed, one at a time, in a unique way (up to equivalence).

9.4 (a) Suppose **y** has weight 4. Since G_{23} is perfect, there is
a unique codeword **x** such that $d(\mathbf{x}, \mathbf{y}) \leqslant 3$, and so
$1 \leqslant w(\mathbf{x}) \leqslant 7$. But every non-zero codeword has weight $\geqslant 7$ and so $w(\mathbf{x}) = 7$, which implies that **x** covers
y. The uniqueness of **x** as a codeword having
distance $\leqslant 3$ from **y** ensures that **x** is the only codeword of weight 7 which covers **y**. Counting in two
ways the number of pairs in the set $\{(\mathbf{x}, \mathbf{y}) \mid \mathbf{x}$ is a
codeword of weight 7, **y** is a vector of weight 4, **x**
covers **y**$\}$ gives $A_7 \cdot \binom{7}{4} = \binom{23}{4} \cdot 1$, whence $A_7 = 253$.

(b) Let $P_1, \ldots, P_{23}$ be points and $B_1, \ldots, B_{253}$ be blocks,
and define $P_i \in B_j$ if and only if the (i, j)th entry of M
is 1.

9.5 (a) Straightforward generalization of the argument of
Exercise 9.4.

(b) Let X be the set of codewords of weight $2t + 1$
beginning with i 1s. Let Y be the set of vectors in
$V(n, 2)$ of weight $t + 1$ beginning with i 1s. As in the
proof of Theorem 9.7, counting in two ways the
number of pairs in the set $\{(\mathbf{x}, \mathbf{y}) \mid \mathbf{x} \in X, \mathbf{y} \in Y, \mathbf{x}$
covers **y**$\}$ gives $\binom{n-i}{t+1-i} = \binom{2t+1-i}{t+1-i} \cdot |X|$,
whence the result.

9.6 We must show that an arbitrary vector $\mathbf{y} = y_1 y_2 \cdots y_{24}$ of
weight 5 in $V(24, 2)$ is covered by a unique codeword of
weight 8 in G_{24}. Certainly there cannot be two such
codewords or their distance apart would be $\leqslant 6$, a contradiction. If $y_{24} = 0$, then since G_{23} is perfect, G_{23} contains
a codeword **x** having distance at most 3 from $y_1 y_2 \cdots y_{23}$.
So **x** has weight 7 or 8; in either case $w(\hat{\mathbf{x}}) = 8$ and $\hat{\mathbf{x}}$
covers **y**. If $y_{24} = 1$, then $y_1 \cdots y_{23}$ is covered by a unique
codeword **x** of weight 7 in G_{23} and then $\hat{\mathbf{x}}$ covers **y**.

9.7 By a now familiar argument, $A_3 \cdot \binom{3}{2} = \binom{2^r - 1}{2}$.

9.8 $A_5 \cdot \binom{5}{3} = \binom{11}{3} \cdot 2^3$.

9.9 (i) Assume $1111111100 \cdots 0 \in G_{24}$. Let G be a generator matrix of G_{24}. Since $d(G_{24}) = 8$ and since G_{24} is
self-dual, it follows by Theorem 8.4 that any 7

columns of G are linearly independent. In particular, the first 7 columns are linearly independent and so by elementary row operations, G may be transformed to a matrix having its first 7 columns as shown. Since $1111111100 \cdots 0$ is orthogonal to every row of G, the eighth column of G must also be as shown.

(ii) Let the rows of G be $\mathbf{r}_1, \mathbf{r}_2, \ldots, \mathbf{r}_{12}$. The set of codewords with one of the given starts is given by adding to $\mathbf{0}$, or to one of $\mathbf{r}_1, \mathbf{r}_2, \ldots, \mathbf{r}_7$, all vectors of the form $\sum_{i=8}^{12} \lambda_i \mathbf{r}_i$, $\lambda_i \in GF(2)$. So for each of the 8 starts, there are 2^5 codewords.

(iii) Immediate, since $d(G_{24}) = 8$, and any two of the chosen 256 codewords differ in at most 2 of the first 8 positions.

(iv) Immediate.

9.10 Shorten N_{15} thrice (cf. Exercise 2.2) to get a $(12, \geq 32, 5)$-code.

9.11 (i) Let the rows of G be $\mathbf{r}_1, \mathbf{r}_2, \ldots, \mathbf{r}_6$. To show that G_2 generates an $[8, 5, 3]$-code, it is enough to show that if $\mathbf{x}$ is any non-zero codeword of C generated by $\mathbf{r}_2, \mathbf{r}_3, \ldots, \mathbf{r}_6$, then $\mathbf{x}$ has at least three 1s in the last 8 positions. If $\mathbf{x}$ had at most two 1s in the last 8 positions, then either $\mathbf{x}$ or $\mathbf{x} + \mathbf{r}_1$ would be a codeword of C having weight ≤ 4, a contradiction.

(ii) If there existed a $[15, 8, 5]$-code, then it could be twice shortened to give a $[13, 6, 5]$-code, contrary to the result of part (i).

(iii) Not immediately, for in this case G_2 would generate a $[7, 4, 3]$-code, and a code with these parameters *does* exist. However, further considerations do lead to a contradiction; see, e.g., van Lint (1982), §4.4.

Chapter 10

10.1 Use Theorem 10.8 with $\mu = 1$, $\nu = 2$.

10.2 In Theorem 10.10, take

$$A_1 = B_1 = \begin{bmatrix} 0 & 1 & 2 \\ 1 & 2 & 0 \\ 2 & 0 & 1 \end{bmatrix}, \qquad A_2 = B_2 = \begin{bmatrix} 0 & 1 & 2 \\ 2 & 0 & 1 \\ 1 & 2 & 0 \end{bmatrix}.$$

10.3 Using Theorem 10.19, a set of three MOLS of order 4 is

$$A_1 = \begin{matrix} 0 & 1 & a & b \\ 1 & 0 & b & a \\ a & b & 0 & 1 \\ b & a & 1 & 0 \end{matrix} \qquad A_2 = \begin{matrix} 0 & a & b & 1 \\ 1 & b & a & 0 \\ a & 0 & 1 & b \\ b & 1 & 0 & a \end{matrix} \qquad A_3 = \begin{matrix} 0 & b & 1 & a \\ 1 & a & 0 & b \\ a & 1 & b & 0 \\ b & 0 & a & 1 \end{matrix}$$

10.4 Ham $(2, q)^{\perp}$ has generator matrix

$$\begin{bmatrix} 0 & 1 & 1 & 1 & & 1 \\ 1 & \lambda_0 & \lambda_1 & \lambda_2 & \cdots & \lambda_{q-1} \end{bmatrix},$$

where $GF(q) = \{\lambda_0, \lambda_1, \ldots, \lambda_{q-1}\}$. Clearly no non-zero linear combination of these two rows can have more than one zero and so Ham $(2, q)^{\perp}$ has minimum distance q. If we list the codewords generated by

$$\begin{bmatrix} 0 & 1 & 1 & 1 & 1 & 1 \\ 1 & 0 & 1 & 2 & 3 & 4 \end{bmatrix}$$

and then apply Theorem 10.20, we get

$$A_1 = \begin{matrix} 0 & 1 & 2 & 3 & 4 \\ 1 & 2 & 3 & 4 & 0 \\ 2 & 3 & 4 & 0 & 1 \\ 3 & 4 & 0 & 1 & 2 \\ 4 & 0 & 1 & 2 & 3 \end{matrix} \qquad A_2 = \begin{matrix} 0 & 1 & 2 & 3 & 4 \\ 2 & 3 & 4 & 0 & 1 \\ 4 & 0 & 1 & 2 & 3 \\ 1 & 2 & 3 & 4 & 0 \\ 3 & 4 & 0 & 1 & 2 \end{matrix}$$

$$A_3 = \begin{matrix} 0 & 1 & 2 & 3 & 4 \\ 3 & 4 & 0 & 1 & 2 \\ 1 & 2 & 3 & 4 & 0 \\ 4 & 0 & 1 & 2 & 3 \\ 2 & 3 & 4 & 0 & 1 \end{matrix} \qquad A_4 = \begin{matrix} 0 & 1 & 2 & 3 & 4 \\ 4 & 0 & 1 & 2 & 3 \\ 3 & 4 & 0 & 1 & 2 \\ 2 & 3 & 4 & 0 & 1 \\ 1 & 2 & 3 & 4 & 0 \end{matrix}$$

10.5

n:	3	4	5	6	7	8	9	10	11	12
$f(n)$:	2	3	4	1	6	7	8	2–9	10	2–11

n:	13	14	15	16	17	18	19	20
$f(n)$:	12	2–13	2–14	15	16	2–17	18	3*–19

* Take three MOLS of order 4 and three MOLS of order 5 and generalize the construction of Theorem 10.10 to get 3 MOLS of order 20.

10.6   The existence of 3 MOLS of order 20 (see previous exercise) gives the existence of a $(5, 400, 4)$-code, by Theorem 10.20. Since this code achieves the Singleton bound, we have $A_{20}(5, 4) = 400$.

Chapter 11

11.1   0204006910.

11.2

$$G = \begin{bmatrix} & & \begin{array}{|cccc} 4 & 7 & 9 & 1 \\ 10 & 8 & 1 & 2 \\ 9 & 7 & 7 & 9 \\ 2 & 1 & 8 & 10 \\ 1 & 9 & 7 & 4 \\ 7 & 6 & 7 & 1 \end{array} \end{bmatrix}$$

with I_6 on the left.

11.3   0000001000, 1005000003.

11.4   Identify the letters $A, B, \ldots, Z$ with the field elements $0, 1, \ldots, 25$ of GF(29). Let H be the parity-check matrix

$$\begin{bmatrix} 1 & 1 & 1 & & 1 \\ 1 & 2 & 3 & & 8 \\ 1 & 2^2 & 3^2 & \cdots & 8^2 \\ 1 & 2^3 & 3^3 & & 8^3 \end{bmatrix}$$

for an $(8, 29^4, 5)$-code over $GF(29)$. Let C be the 26-ary code obtained by taking only those codewords consisting of symbols $0, 1, \ldots, 25$, i.e.

$$C = \left\{ x_1 x_2 \cdots x_8 \mid x_i \in \{0, 1, \ldots, 25\}, \right.$$

$$\left. \sum_{i=1}^{8} i^j x_i \equiv 0 \ (\text{mod } 29), j = 0, 1, 2, 3 \right\}.$$

A probabilistic estimate for the number of codewords in C is $29^4 \times \left(\frac{26}{29}\right)^8 \approx 295, 253$ (it happens that this is a remarkably good estimate).

Alternatively we could base our code on 26 of the elements of $GF(27)$. This would give us more codewords, but the arithmetic involved in the decoding would be less straightforward.

11.5 $\sigma(\theta) = \prod_{i=1}^{e} (1 - X_i\theta) \Rightarrow \sigma'(\theta) = -\sum_{t=1}^{e} X_t \prod_{\substack{i=1 \\ i \neq t}}^{e} (1 - X_i\theta)$

$$\Rightarrow \sigma'(X_j^{-1}) = -X_j \prod_{\substack{i=1 \\ i \neq j}}^{e} (1 - X_i X_j^{-1}).$$

The result now follows from equation (11.10).

11.6 $H = \begin{bmatrix} 3 & 7 & 6 & 1 & 8 & 9 & 4 & 5 & 2 & 1 & 0 \\ 5 & 4 & 3 & 2 & 7 & 6 & 5 & 4 & 3 & 2 & 1 \end{bmatrix}$.

There exists a codeword $(x_1, \ldots, x_{11})$ of weight 2 with non-zero entries x_i and x_j if and only if $H_i = -(x_j/x_i)H_j$, where H_i denotes the ith column of H. In order to determine which columns of H are scalar multiples of others, calculate the ratios h_1/h_2 for each column

$$\begin{bmatrix} h_1 \\ h_2 \end{bmatrix}.$$

They are 5, 10, 2, 6, 9, 7, 3, 4, 8, 6, 0. It follows that a double-error vector will go undetected if and only if it is of the form $(0, 0, 0, \lambda, 0, 0, 0, 0, 0, -\lambda, 0)$ for some $\lambda \in \{1, 2, \ldots, 10\}$.

Chapter 12

12.1 (i) No, No (not linear), (ii) No, No, (iii) No, Yes, (iv) Yes, provided the alphabet is a field, (v) Yes, (vi) No, No, (vii) Yes.

12.2

$\cdot$	0	1	x	$1+x$	$1+x$ has no inverse
0	0	0	0	0	
1	0	1	x	$1+x$	
x	0	x	1	$1+x$	
$1+x$	0	$1+x$	$1+x$	0	

12.3 Just imitate the proof of Theorem 3.5.

12.4 If $f(x)$ had an even number of non-zero coefficients, then we would have $f(1) = 0$ and so $x - 1$ would be a factor of $f(x)$.

12.5   Because $p(x) = f(x)g(x) \Rightarrow \deg p(x) = \deg f(x) + \deg g(x)$
$$\Rightarrow \text{either } \deg f(x) \leqslant \tfrac{1}{2}\deg p(x) \text{ or}$$
$$\deg g(x) \leqslant \tfrac{1}{2}\deg p(x).$$

12.6 x, $1 + x$, $1 + x + x^2$, $1 + x + x^3$, $1 + x^2 + x^3$, $1 + x + x^4$, $1 + x^3 + x^4$, $1 + x + x^2 + x^3 + x^4$. (Using Lemma 12.3 and Exercise 12.4, it easily follows that the irreducible polynomials of degrees 2, 3 and 4 are precisely those with constant coefficient 1 and with an odd number of non-zero coefficients, with the exception of $(1 + x + x^2)^2 = 1 + x^2 + x^4$). For example, $F[x]/(1 + x + x^3)$ is a field of order 8.

12.7   (i)   By Exercise 3.12, $(x^p - 1) = (x - 1)^p$.
       (ii) From Fermat's theorem (Exercise 3.8) and Lemma 12.3(i), it follows that $x^{p-1} = (x - 1)(x - 2) \cdots (x - (p - 1))$.

12.8   By Lemma 12.3(i), $x^5 - 1 = (x - 1)(x^4 + x^3 + x^2 + x + 1)$, and the second factor is irreducible by Exercise 12.6. So the only cyclic codes are $\{\mathbf{0}\}$, $\langle x - 1 \rangle$ (the even weight code), $\langle x^4 + x^3 + x^2 + x + 1 \rangle$ (the repetition code), and the whole of $V(5, 2)$.

12.9   Yes, $(x - 1)g(x)$.

12.10 2^t. (In a factor of $x^n - 1$, each of the t distinct irreducible factors may or may not be present).

12.11 $\langle 1 \rangle$ = whole space
       $\langle x - 1 \rangle$ = even weight code E_7
       $\left.\begin{array}{l} \langle x^3 + x + 1 \rangle \\ \langle x^3 + x^2 + 1 \rangle \end{array}\right\}$both are Hamming codes Ham $(3, 2)$
       $\left.\begin{array}{l} \langle (x - 1)(x^3 + x + 1) \rangle \\ \langle (x - 1)(x^3 + x^2 + 1) \rangle \end{array}\right\}$both are even weight subcodes of Ham $(3, 2)$ (alternatively, both are duals of Ham $(3, 2)$)
       $\langle (x^3 + x + 1)(x^3 + x^2 + 1) \rangle$ = repetition code of length 7
       $\langle x^7 - 1 \rangle = \{\mathbf{0}\}$.

12.12 $x^8 - 1 = (x^4 - 1)(x^4 + 1) = (x - 1)(x + 1)(x^2 + 1)(x^2 + x + 2)(x^2 + 2x + 2)$, 32.

12.13 Straightforward application of Theorem 12.15.

12.14 Not in general; Yes, $C^\perp$ is obtained from $\langle h(x) \rangle$ by writing the codewords backwards.

12.15 Let $g(x)$ be the generator polynomial of C. Then $g(x)$ is a divisor of $(x-1)(x^{n-1}+\cdots+x+1)$. If $g(x)$ is a multiple of $x-1$, then so is every codeword, and so every codeword has even weight. So if there exists a codeword of odd weight, then $x^{n-1}+\cdots+x+1$ must be a multiple of $g(x)$, i.e. $\mathbf{1}\in C$. The reverse implication is immediate since $w(\mathbf{1})$ is odd.

12.16 Let $\mathbf{g}_1,\ldots,\mathbf{g}_k$ denote the rows of G. Let $\tilde{\mathbf{x}}$ denote a cyclic shift of $\mathbf{x}$. If $\mathbf{x}=\sum\lambda_i\mathbf{g}_i\in C$, then $\tilde{\mathbf{x}}=\sum\lambda_i\tilde{\mathbf{g}}_i\in C$.

12.17 Check that $2^0,2^1,\ldots,2^9$ are precisely the distinct nonzero elements of $GF(11)$. Hence the code of Example 7.12 is equivalent to the code C with parity-check matrix

$$\begin{bmatrix}1 & 1 & 1 & & 1\\ 2^0 & 2^1 & 2^2 & \cdots & 2^9\end{bmatrix}.$$

Now

$$(2^9,2^0,2^1,\ldots,2^8)=2^9(2^0,2^1,\ldots,2^9)$$

and so $C^\perp$ is cyclic by Exercise 12.16. Therefore C is cyclic by Theorem 12.15(ii). The result for Example 11.3 follows similarly.

12.18 The subcode D of G_{23} consisting of codewords of even weight is $\langle(x-1)g_1(x)\rangle$. Thus $D^\perp=\langle\bar{g}_2(x)\rangle=\langle g_1(x)\rangle$ and so $D\subseteq D^\perp$. Hence $\mathbf{u}\cdot\mathbf{v}=0$ if $\mathbf{u}$ and $\mathbf{v}$ are codewords of even weight. Since $\mathbf{1}\in G_{23}$, any codeword of odd weight is of the form $\mathbf{u}+\mathbf{1}$ for some codeword $\mathbf{u}$ of even weight. If $\mathbf{u}+\mathbf{1},\mathbf{v}+\mathbf{1}$ are codewords of odd weight, then $(\mathbf{u}+\mathbf{1})\cdot(\mathbf{v}+\mathbf{1})=\mathbf{u}\cdot\mathbf{v}+\mathbf{1}\cdot\mathbf{v}+\mathbf{u}\cdot\mathbf{1}+\mathbf{1}\cdot\mathbf{1}=0+0+0+1=1$. Also if $\mathbf{u}+\mathbf{1}$ has odd weight and $\mathbf{v}$ has even weight, then $(\mathbf{u}+\mathbf{1})\cdot\mathbf{v}=\mathbf{u}\cdot\mathbf{v}+\mathbf{1}\cdot\mathbf{v}=0+0=0$. Now let $\mathbf{x},\mathbf{y}$ be any codewords of G_{23} and let $\hat{\mathbf{x}},\hat{\mathbf{y}}$ be the corresponding codewords of G_{24}. Then $\hat{\mathbf{x}}\cdot\hat{\mathbf{y}}=\mathbf{x}\cdot\mathbf{y}+x_{24}y_{24}=0$, since $\mathbf{x}\cdot\mathbf{y}=1\Leftrightarrow\mathbf{x},\mathbf{y}$ both have odd weight $\Leftrightarrow x_{24}=y_{24}=1$. So $G_{24}\subseteq G_{24}^\perp$ and since dim$(G_{24})=$ dim$(G_{24}^\perp)=12$, it follows that $G_{24}=G_{24}^\perp$.

12.19 x^4+x+1 is a generator polynomial for Ham$(4,2)$. Dividing $x^{15}-1$ by x^4+x+1 (e.g. by long division) gives $h(x)=x^{11}+x^8+x^7+x^5+x^3+x^2+x+1$.

12.20 Ham$(r,2)$ is a $[2^r-1,2^r-r-1,3]$-code. By Exercise 12.9, $\langle(x-1)g(x)\rangle$ is the subcode of codewords of even weight. This subcode must have dimension $2^r - r - 2$ and minimum distance 4.

12.21 It is enough to show that no vector of the form

$$(x^i + x^{i+1}) + (x^j + x^{j+1}) = (x + 1)(x^i + x^j)$$

is a codeword of $\langle (x + 1)g(x) \rangle$ (then all vectors of the form 0, x^i, and $x^i + x^{i+1}$ will be coset leaders). But $(x + 1)(x^i + x^j)$ is a codeword $\Rightarrow (x + 1)(x^i + x^j)$ is a multiple of $(x + 1)g(x) \Rightarrow x^i + x^j$ is a multiple of $g(x) \Rightarrow x^i + x^j \in \langle g(x) \rangle$, contradicting $d(\langle g(x) \rangle) = 3$.

12.22 (van Lint 1982, solution to Exercise 6.11.7). Show that every non-zero codeword of C has exactly one zero entry. Show also that there is exactly one codeword $\mathbf{c} = c_0 c_1 \cdots c_q$ such that $c_0 = c_{(q+1)/2} = 1$ [Consider the q^2 ordered pairs $(c_0, c_{(q+1)/2})$ as $\mathbf{c}$ runs over all codewords of C]. If C were cyclic, then a cyclic shift of $\mathbf{c}$ through $(q + 1)/2$ positions would yield the same codeword $\mathbf{c}$, but this is not possible if $\mathbf{c}$ contains only one zero entry. Thus C is not cyclic and so Ham $(2, q)$, being the dual code of C, is not cyclic by Theorem 12.15(ii).

Chapter 13

13.1 The mapping $\mathbf{x} \to \mathbf{x} + \mathbf{1}$ gives a one-to-one correspondence between the set of codewords of weight i and the set of codewords of weight $n - i$.

13.2 (b) $C^\perp$ is generated by

$$\begin{bmatrix} 1 & 0 & 0 & 1 & 0 \\ 1 & 1 & 1 & 0 & 1 \end{bmatrix},$$

and so $C^\perp = \{00000, 10010, 11101, 01111\}$. Hence $W_{C^\perp}(z) = 1 + z^2 + 2z^4$. So

$$W_C(z) = \tfrac{1}{4}(1 + z)^5 W_{C^\perp}\left(\frac{1 - z}{1 + z}\right)$$

$$= \tfrac{1}{4}[(1 + z)^5 + (1 + z)^3(1 - z)^2 + 2(1 + z)(1 - z)^4]$$

$$= 1 + 3z^2 + 3z^3 + z^5.$$

13.3 $C^\perp$ is generated by

$$\begin{bmatrix} 0 & 0 & 1 & 1 & 1 & 1 & 1 & 1 & 0 \\ 1 & 1 & 0 & 0 & 1 & 1 & 1 & 0 & 1 \end{bmatrix},$$

and so $W_{C^\perp}(z) = 1 + 3z^6$. Hence $W_C(z) = \frac{1}{4}[(1 + z)^9 + 3(1 + z)^3(1 - z)^6]$, whence $A_0 = 1$, $A_1 = 0$, $A_2 = \frac{1}{4}[36 + 3(3 + 15 - 18)] = 9$, $A_3 = 27$. The sum of all the rows of the generator matrix of C is **1**. By Exercise 13.1, $A_i = A_{9-i}$, and so $2(A_0 + A_1 + A_2 + A_3 + A_4) = 2^7$, which gives $A_4 = 27$. Hence

$$W_C(z) = 1 + 9z^2 + 27z^3 + 27z^4 + 27z^5 + 27z^6 + 9z^7 + z^9.$$

13.4 Adding an overall parity check increases each odd weight by 1 and leaves each even weight unchanged. So $W_{\hat{C}}(z) = 1 + 14z^4 + z^8$.

13.5 Let C be Ham $(r, 2)$. Then by Theorem 13.10, $W_{C^\perp}(z) = 1 + (2^r - 1)z^{2^{r-1}} = 1 + nz^{(n+1)/2}$. So

$$W_C(z) = \frac{1}{2^r}[(1 + z)^n + n(1 - z)^{(n+1)/2}(1 + z)^{(n-1)/2}]$$

$$= \frac{1}{2^r}[(1 + z)^n + n(1 - z^2)^{(n-1)/2}(1 - z)].$$

13.6 $W_C(z) = \frac{1}{16}[(1 + z)^{15} + 15(1 - z^2)^7(1 - z)].$ (1)
$A_0 = 1$, $A_1 = A_2 = 0$ (either from (1) or because we know $d(C) = 3$), $A_3 = 35$, $A_4 = 105$.

13.7 The coefficient of z^i in the right-hand side is $\frac{1}{2}A_i(1 + (-1)^i) = A_i$ if i is even, 0 if i is odd.

13.8 If $W_C(z) = \sum A_i z^i$, then

$$W_{\hat{C}}(z) = \sum_{i \text{ even}} (A_i + A_{i-1})z^i$$

$$= \sum_{i \text{ even}} A_i z^i + z \sum_{j \text{ odd}} A_j z^j$$

$$= \frac{1}{2}[W_C(z) + W_C(-z)] + \frac{1}{2}z[W_C(z) - W_C(-z)].$$

13.9 From equation (13.12),

$$P_{\text{undetec}}(C) = (1 - p)^n \frac{1}{2^{n-k}}\left(1 + \frac{p}{1 - p}\right)^n$$

$$\times W_{C^\perp}\left(\left(1 - \frac{p}{1 - p}\right)\Big/\left(1 + \frac{p}{1 - p}\right)\right)$$

$$- (1 - p)^n$$

$$= \frac{1}{2^{n-k}} W_{C^\perp}(1 - 2p) - (1 - p)^n.$$

13.10 By Lemmas 1, 3, and 4 in the proof of Theorem 9.3, G_{24} is self-dual, $A_i \neq 0$ only if i is divisible by 4, and $A_4 = 0$. Since $1 \in G_{24}$, it follows from Exercise 13.1 that $A_{20} = 0$ and that $A_{16} = A_8$. So $W_C(z) = 1 + A_8 z^8 + A_{12} z^{12} + A_8 z^{16} + z^{24}$. Applying the MacWilliams identity and equating coefficients of $W_C(z)$ and $W_{C^\perp}(z)$ (since C is self-dual) gives: $2 + 2A_8 + A_{12} = 2^{12}$ (constant coefficients) $0 = 0$ (coefficients of z) and $138 + 10A_8 - 3A_{12} = 0$ (coefficients of z^2). Solving these gives $A_8 = 759$, $A_{12} = 2576$.

13.11 G_{24} is self-dual by Exercise 12.18. By Lemma 12.19, codewords of G_{23} of even weight have weight divisible by 4. Since $1 (=g_1(x)g_2(x)) \in G_{23}$, it follows by Exercise 13.1 that any odd weight of a codeword of G_{23} is congruent to 3 (mod 4). Consequently, all codewords of G_{24} have weight divisible by 4. Also $A_4 = 0$, since $d(G_{24}) = 8$. The result now follows exactly as in Exercise 13.10.

13.12 By Exercise 13.10 (or 13.11) the only A_is in $W_{G_{23}}(z)$ which can be non-zero are A_0, A_7, A_8, A_{11}, A_{12}, A_{15}, A_{16} and A_{23}. Also $A_7 + A_8 = 759$ and $A_{11} + A_{12} = 2576$. By Exercise 9.4(a), $A_7 = 253$ and so $A_8 = 506$. Since $1 \in G_{23}$, we have $A_{11} = A_{12} = 1288$, $A_{15} = 506$, and $A_{16} = 253$. So

$$W_{G_{23}}(z) = 1 + 253z^7 + 506z^8 + 1288z^{11}$$
$$+ 1288z^{12} + 506z^{15} + 253z^{16} + z^{23}.$$

Chapter 14

14.1 No; Exercises 9.9 and 9.11 show that $A_2(15, 5) \geqslant 256$, $B_2(15, 5) = 128$.

14.2 (i) $V(n, q)$ is an $[n, n, 1]$-code.
      (ii) $C = \{x_1 x_2 \cdots x_n \mid x_1 + x_2 + \cdots + x_n = 0\}$ is an $[n, n-1, 2]$-code. Since there cannot exist an $[n, n, 2]$-code, we have $B_q(n, 2) = q^{n-1}$.

14.3 By Theorem 14.4, there exists an $[n, n-r, 3]$-code over

$$GF(q) \Leftrightarrow n \leqslant (q^r - 1)/(q-1)$$
$$\Leftrightarrow r \geqslant \log_q \{n(q-1)+1\}$$
$$\Leftrightarrow n - r \leqslant n - \log_q \{n(q-1)+1\}.$$

So $B_q(n, 3) = q^{\lfloor n - \log_q \{n(q-1)+1\} \rfloor}$.

14.4 Let t be the number of planes in which a given line L lies. Counting in two ways the number of members of the set $\{(P, \pi) \mid P$ is a point not on L, π is a plane containing both L and $P\}$ gives $q^3 + q^2 + q + 1 - (q+1) = t[q^2 + q + 1 - (q+1)]$, whence $t = q + 1$.

14.5 The Golay code G_{11} is a ternary $[11, 6, 5]$-code, showing that $\max_4 (5, 3) \geqslant 11$. If $\max_4 (5, 3)$ were $\geqslant 12$, then there would exist a ternary $[12, 7, 5]$-code, contradicting the sphere-packing bound.

14.6 Use Theorem 14.18. Since 2 is a non-square in $GF(5)$, the 4×26 matrix whose columns are $(0, 0, 0, 1)^T$ and $(x, y, 1, x^2 - 2y^2)^T$, for $(x, y) \in V(2, 5)$, is the parity-check matrix of a $[26, 22, 4]$-code.

14.7 (i) By Theorem 14.16, a plane can contain $q + 2$ points of a cap.

   (ii) By Exercise 3.14, if q is even, then every element of $GF(q)$ is a square. [*Remark:* a version of Theorem 14.18 does hold for q even, with an elliptic quadric specified in a different way].

14.8 Let H be the parity-check matrix whose columns form the $(q^2 + 1)$-cap defined in (14.19). Label the column $(0001)^T$ by ∞ and each column $(x, y, 1, x^2 - by^2)^T$ by (x, y). A decoding algorithm is the following. Calculate the syndrome $s = yH^T = s_1 s_2 s_3 s_4$. If $s = 0$, assume no errors. If $s \neq 0$, calculate $\theta = s_3 s_4 - s_1^2 + bs_2^2$. If $\theta = 0$ and $s_3 = 0$, assume an error of magnitude s_4 in position ∞. If $\theta = 0$ and $s_3 \neq 0$, assume an error of magnitude s_3 in position $(s_1/s_3, s_2/s_3)$. If $\theta \neq 0$, then there are $\geqslant 2$ errors.

14.9 If $\{x_1, x_2, \ldots, x_{10}\}$ is a 10-cap in $PG(3, 3)$ then the set

$$\{(x_1 \mid 0), (x_2 \mid 0), \ldots, (x_{10} \mid 0), (x_1 \mid 1), (x_2 \mid 1), \ldots,$$
$$(x_{10} \mid 1)\}$$

is a 20-cap in $PG(4, 3)$.

14.10 For a given t-space, the number of ways of choosing an extra point of $PG(m, q)$ to generate a $(t + 1)$-space is

$$\frac{q^{m+1} - 1}{q - 1} - \frac{q^{t+1} - 1}{q - 1}.$$

Many of these extra points generate the same $(t + 1)$-space, and so we must divide by

$$\frac{q^{t+2} - 1}{q - 1} - \frac{q^{t+1} - 1}{q - 1},$$

the number of points in such a $(t + 1)$-space not lying in the given t-space.
(i) 40, (ii) 13, (iii) 4.

14.11 (Bruen and Hirschfeld 1978). Suppose K is a cap in $PG(5, 3)$. We shall show that $|K| \leq 56$. We may assume some plane π meets K in four points, for otherwise $|K| \leq 42$ (two points on some line L plus at most one further point on each of the 40 planes through L). Similarly, we may assume some 3-space contains at least 8 points of K, for otherwise $|K| \leq 4 + 3 \cdot 13 = 43$. Finally, since $\max_3 (5, 3) = 20$, we have $|K| \leq 8 + 4(20 - 8) = 56$.

14.12 $B_3(n, 4) = 3^{n-4}$ for $5 \leq n \leq 10$, 3^{n-5} for $11 \leq n \leq 20$, 3^{n-6} for $21 \leq n \leq 56$, 3^{n-7} for $57 \leq n \leq 112$. (It is not known whether $B_3(113, 4) = 3^{106}$ or 3^{105}.)

Chapter 15

15.1 $[I_3 \,|\, A]$, $[I_5 \,|\, A^T]$.

15.2 Suppose, for a contradiction, that $\max_{q+2-r} (q + 2 - r, q) \geq q + 2$. Then there exists a $[q + 2, r, q + 3 - r]$-code whose dual is a $[q + 2, q + 2 - r, r + 1]$-code, contradicting $\max_r (r, q) = q + 1$.

15.3 $\max_{q-1} (q - 1, q) \geq q + 2$ by Corollary 15.9. If there existed a $[q + 3, 4, q]$-code over $GF(q)$, then its dual would be a $[q + 3, q - 1, 5]$-code, contrary to $\max_4 (4, q) = q + 1$.

15.4
$$H = \begin{bmatrix} & & 1 & 1 & 1 \\ & & 1 & a_2 & a_2^2 \\ & I_7 & 1 & a_3 & a_3^2 \\ & & \vdots & \vdots & \vdots \\ & & 1 & a_6 & a_6^2 \end{bmatrix}.$$

15.5 Let $H = [A \mid I]$ be a standard form parity-check matrix for an $[n, n - r, r + 1]$-code with $n = \max_r (r, q)$. Deleting the last row and last column of H gives a matrix whose columns form an $(n - 1, r - 1)$-set in $V(r - 1, q)$ and so $n - 1 \leqslant \max_{r-1} (r - 1, q)$.

15.6 Let C be an $[8, 3, 6]$-code over $GF(7)$. By Corollary 15.7, $C^\perp$ is an $[8, 5, 4]$-code. Let $W_C(z) = \sum A_i z^i$ and $W_{C^\perp}(z) = \sum B_i z^i$. By Theorem 13.6,

$$7^3 \left(1 + \sum_{i=4}^{8} B_i z^i \right) = (1 + 6z)^8 + A_6(1 - z)^6(1 + 6z)^2 \\ + A_7(1 - z)^7(1 + 6z) + A_8(1 - z)^8 \quad (1)$$

Equating coefficients of 1, z and z^2 and solving for A_6, A_7 and A_8 gives $W_C(z) = 1 + 168z^6 + 48z^7 + 126z^8$. $W_{C^\perp}(z)$ is now easily obtained directly from (1).

15.7 For $2 \leqslant k \leqslant 11$,

$$\begin{bmatrix} 1 & 1 & \cdots & 1 & 1 & 0 \\ 1 & 2 & \cdots & 10 & 0 & 0 \\ \vdots & \vdots & & \vdots & \vdots & \vdots \\ 1 & 2^{k-1} & \cdots & 10^{k-1} & 0 & 1 \end{bmatrix}$$

generates a $[12, k, 13 - k]$-code.
For $k \geqslant 11$, $\begin{bmatrix} & 1 \\ I_k & 1 \\ & \vdots \\ & 1 \end{bmatrix}$ generates a $[k + 1, k, 2]$-code.

Chapter 16

16.1 If $x \not\equiv 0 \,(\mathrm{mod}\, p)$, then $x^{r(p-1)(q-1)+1} = (x^{p-1})^{r(q-1)} x \equiv x$ $(\mathrm{mod}\, p)$ by Fermat's theorem. If $x \equiv 0 \,(\mathrm{mod}\, p)$, then $x^{r(p-1)(q-1)+1} \equiv x \,(\mathrm{mod}\, p)$ holds trivially. So p is a factor of $x^{r(p-1)(q-1)+1} - x$ for any integer x. Similarly q is also a factor for any integer x. Since p and q are distinct prime numbers, pq is a factor of $x^{r(p-1)(q-1)+1} - x$ for any x.

16.2 LEAVING TOMORROW.

16.3 When the subscriber R (of the text) has encrypted a message he is to send to S (using S's encryption algorithm) he signs it with a further message z which he sends in the form $z^t \pmod{n}$ (i.e. via R's own *de*crypting algorithm). The receiver S verifies the signature by calculating $(z^t)^s \equiv z \pmod{n}$. Only R could have sent the message, since only R knows t.

16.4 B, C and D are uniquely decodable. B and C are prefix-free. Average word-lengths of B, C and D are 2, 1.75 and 1.875, respectively. [*Remark:* It is a consequence of Shannon's 'source coding theorem' (see, e.g., Jones 1979) that the 'source entropy', $-\sum_{i=1}^{4} p_i \log_2 p_i$ ($=1.75$ here), gives the smallest possible average word length. So the above code C here is best possible.]

16.5 THE END.

Bibliography

At the end of each entry the number in square brackets gives the chapter which refers to this entry.

Anderson, I. (1974). *A first course in combinatorial mathematics.* Clarendon Press, Oxford. [2, 16]

Assmus, E. F. and Mattson, H. F. (1967). On tactical configurations and error-correcting codes. *J. Comb. Theory* **2**, 243–57. [9]

—— (1969). New 5-designs. *J. Comb. Theory* **6**, 122–51. [9]

—— (1974). Coding and combinatorics. *SIAM Review* **16**, 349–88. [9]

Barlotti, A. (1955). Un estensione del teorema di Segre-Kustaanheimo. *Boll. Un. Mat. Ital.* **10**, 96–8. [14]

Beker, H. and Piper, F. (1982). *Cipher Systems.* Van Nostrand Reinhold, London. [16]

Berlekamp, E. R. (1968). *Algebraic coding theory.* McGraw-Hill, New York. [Pref., 11]

—— (1972). Decoding the Golay code. *JPL Technical Report* 32–1526, Vol. IX, 81–5. [9]

Best, M. R. (1980). Binary codes with a minimum distance of four. *IEEE Trans. Info. Theory* **26**, 738–42. [16]

—— (1982). A contribution to the nonexistence of perfect codes. Ph.D. dissertation, University of Amsterdam. [9]

—— (1983). Perfect codes hardly exist. *IEEE Trans. Info. Theory* **29**, 349–51. [9]

—— and Brouwer, A. E. (1977). The triply shortened Hamming code is optimal. *Discrete Math.* **17**, 235–45. [8]

Best, M. R., Brouwer, A. E., MacWilliams, F. J., Odlyzko, A. M., and Sloane, N. J. A. (1978). Bounds for binary codes of length less than 25. *IEEE Trans. Info. Theory* **24**, 81–92. [16]

Blahut, R. E. (1983). *Theory and practice of error control codes.* Addison-Wesley, Reading, Mass. [Pref., 11, 12, 16]

Blake, I. F. and Mullin, R. C. (1976). *An introduction to algebraic and combinatorial coding theory.* Academic Press, New York. [Pref.]

Blokhuis, A. and Lam, C. W. H. (1984). More coverings by rook domains. *J. Comb. Theory,* Ser. A **36**, 240–4. [8]

Bose, R. C. (1947). Mathematical theory of the symmetrical factorial design. *Sankhyā* **8**, 107–166. [14]

—— and Ray-Chaudhuri, D. K. (1960). On a class of error-correcting binary group codes. *Info. and Control* **3**, 68–79. [11, 12]

——, Shrikhande, S. S., and Parker, E. T. (1960). Further results on the construction of mutually orthogonal Latin squares and the falsity of Euler's conjecture. *Canad. J. Math.* **12**, 189–203. [10]

Brinn, L. W. (1984). Algebraic coding theory in the undergraduate curriculum. *American Math. Monthly*, 509–13. [Pref.]

Brown, D. A. H. (1974). Some error correcting codes for certain transposition and transcription errors in decimal integers. *Computer Journal* **17**, 9–12. [11]

Bruen, A. A. and Hirschfeld, J. W. P. (1978). Application of line geometry over finite fields. II. The Hermitian surface. *Geom. Dedicata* **7**, 333–53. [14]

Bush, K. A. (1952). Orthogonal arrays of index unity. *Ann. Math. Stat.* **23**, 426–34. [15]

Cameron, P. J. and van Lint, J. H. (1980). *Graphs, codes and designs.* London Math. Soc. Lecture Note Series, Vol. 43. Cambridge Univ. Press, Cambridge. [Pref.]

Casse, L. R. A. (1969). A solution to Beniamino Segre's 'Problem $I_{r,q}$' for q even. *Atti. Accad. Naz. Lincei Rend.* **46**, 13–20. [15]

Delsarte, P. and Goethals, J.-M. (1975). Unrestricted codes with the Golay parameters are unique. *Discrete Math.* **12**, 211–24. [9]

Dénes, J. and Keedwell, A. D. (1974). *Latin squares and their applications.* Academic Press, New York. [10]

Diffie, W. and Hellman, M. E. (1976). New directions in cryptography. *IEEE Trans. Info. Theory* **22**, 644–54. [16]

Dornhoff, L. L. and Hohn, F. E. (1978). *Applied Modern Algebra.* Macmillan, New York. [16]

Fenton, N. E. and Vámos, P. (1982). Matroid interpretation of maximal k-arcs in projective spaces. *Rend. Mat.* (7) **2**, 573–80. [14, 15]

Fernandes, H. and Rechtschaffen, E. (1983). The football pool problem for 7 and 8 matches. *J. Comb. Theory,* Series A **35**, 109–14. [8]

Games, R. A. (1983). The packing problem for projective geometries over $GF(3)$ with dimension greater than five. *J. Comb. Theory,* Series A **35**, 126–44. [14]

Gardner, M. (1977). Mathematical games. *Scientific American*, August, 120–4. [16]

Gibson, I. B. and Blake, I. F. (1978). Decoding the binary Golay code with miracle octad generators. *IEEE Trans. Info. Theory* **24**, 261–4. [9]

Gilbert, E. N. (1952). A comparison of signalling alphabets. *Bell Syst. Tech. J.* **31**, 504–22. [8]

Goethals, J.-M. (1971). On the Golay perfect binary code. *J. Comb. Theory* **11**, 178–86. [9]

—— (1977). The extended Nadler code is unique. *IEEE Trans. Info. Theory* **23**, 132–5. [9]

Golay, M. J. E. (1949). Notes on digital coding. *Proc. IEEE* **37**, 657. [8, 9]

—— (1954). Binary coding. *Trans IRE PGIT* **4**, 23–8. [16]

—— (1958). Notes on the penny-weighing problem, lossless symbol coding with nonprimes, etc. *IEEE Trans. Info. Theory* **4**, 103–9. [9]

Golomb, S. W. and Posner, E. C. (1964). Rook domains, Latin squares, affine planes, and error-distribution codes. *IEEE Trans. Info. Theory* **10**, 196–208. [9]

Goppa, V. D. (1970). A new class of linear error-correcting codes. *Problems of Info. Transmission* **6** (3), 207–12. [11]

Hall, M. (1980). *Combinatorial theory.* Wiley, New York. [2]

Hamming, R. W. (1950). Error detecting and error correcting codes. *Bell Syst. Tech. J.* **29**, 147–60. [8]

—— (1980). *Coding and information theory.* Prentice-Hall, New Jersey. [16]

Hardy, G. H. (1940). *A mathematician's apology.* Cambridge University Press. [11]

Helgert, H. J. and Stinaff, R. D. (1973). Minimum distance bounds for binary linear codes. *IEEE Trans. Info. Theory* **19**, 344–56. [14]

Hill, R. (1973). On the largest size of cap in $S_{5,3}$. *Atti Accad. Naz. Lincei Rend.* **54**, 378–84. [14]

—— (1978). Caps and codes. *Discrete Math.* **22**, 111–37. [14]

Hirschfeld, J. W. P. (1979). *Projective geometries over finite fields.* Oxford University Press. [14]

—— (1983). Maximum sets in finite projective spaces. In *Surveys in combinatorics,* LMS Lecture Note Series 82, edited by E. K. Lloyd. Cambridge University Press, 55–76. [14]

Hocquenghem, A. (1959). Codes correcteurs d'erreurs. *Chiffres* (Paris) **2**, 147–58. [11]

Jones, D. S. (1979). *Elementary information theory.* Clarendon Press, Oxford. [16]

Jurick, R. R. (1968). An algorithm for determining the largest maximally independent set of vectors from an r-dimensional space over a Galois field of n elements. Tech. Rep. ASD–TR–68–40, Air Force Systems Command, Wright–Patterson Air Force Base, Ohio. [15]

Kamps, H. J. L. and van Lint, J. H. (1967). The football pool problem for 5 matches. *J. Comb. Theory* **3**, 315–25. [8]

Levenshtein, V. I. (1964). The application of Hadamard matrices to a problem in coding. *Problems of Cybernetics* **5**, 166–84. [16]

Levinson, N. (1970). Coding theory: a counterexample to G. H. Hardy's conception of applied mathematics. *Amer. Math. Monthly* **77**, 249–58. [11]

Lidl, R. and Niederreiter, H. (1983). *Finite fields.* Addison-Wesley, and (1984) Cambridge University Press. [3]

Lin, S. and Costello, D. J. (1983). *Error control coding: fundamentals and applications.* Prentice-Hall, New Jersey. [Pref., 12]

Lindstrom, B. (1969). On group and nongroup perfect codes in *q* symbols. *Math. Scand.* **25**, 149–58. [9]

van Lint, J. H. (1975). A survey of perfect codes. *Rocky Mountain J. of Mathematics* **5**, 199–224. [9]

—— (1982). *Introduction to coding theory.* Springer-Verlag, New York. [Pref., 6, 9, 16]

Lloyd, S. P. (1957). Binary block coding. *Bell Syst. Tech. J.* **36**, 517–35. [9]

McEliece, R. J. (1977). *The theory of information and coding.* Addison-Wesley, Reading, Mass. [Pref., 6, 16]

McEliece, R. J., Rodemich, E. R., Rumsey, H. and Welch, L. R. (1977). New upper bounds on the rate of a code via the Delsarte–MacWilliams inequalities. *IEEE Trans. Info. Theory* **23**, 157–66. [16]

Mackenzie, C. and Seberry, J. (1984). Maximal ternary codes and Plotkin's bound. *Ars Combinatoria* **17A**, 251–70. [2]

MacWilliams, F. J. (1963). A theorem on the distribution of weights in a systematic code. *Bell Syst. Tech. J.* **42**, 79–94. [13]

—— and Sloane, N. J. A. (1977). *The theory of error-correcting codes.* North-Holland, Amsterdam. [Pref., 2, 8, 9, 11–15, 16]

Magliveras, S. S. and Leavitt, D. W. (1983). Simple six-designs exist. *Congressus Numerantium* **40**, 195–205. [9]

Maneri, C. and Silverman, R. (1966). A vector space packing problem. *J. of Algebra* **4**, 321–30. [15]

Massey, J. L. (1969). Shift-register synthesis and BCH decoding. *IEEE Trans. Info. Theory* **15**, 122–27. [11]

Nadler, M. (1962). A 32-point *n* = 12, *d* = 5 code. *IEEE Trans. Info. Theory* **8**, 58. [9]

Nordstrom, A. W. and Robinson, J. P. (1967). An optimum non-linear code. *Info. and Control* **11**, 613–16. [2]

Pellegrino, G. (1970). Sul massimo ordine delle calotte in $S_{4,3}$. *Matematiche* (Catania) **25**, 1–9. [14]

Peterson, W. W. and Weldon, E. J. (1972). *Error-correcting codes,* 2nd ed. MIT Press, Cambridge, Mass. [Pref., 11, 16]

Phelps, K. T. (1983). A combinatorial construction of perfect codes. *SIAM J. Alg. Disc. Math.* **4**, 398–403. [9]

Pless, V. (1968). On the uniqueness of the Golay codes. *J. Comb. Theory* **5**, 215–28. [9]

—— (1982). *Introduction to the theory of error-correcting codes.* Wiley, New York. [Pref., 9]

Plotkin, M. (1960). Binary codes with specified minimum distance. *IEEE Trans. Info. Theory* **6**, 445–450. [2]

Prange, E. (1957). Cyclic error-correcting codes in two symbols. Technical Note TN–57–103, Air Force Cambridge Research Labs., Bedford, Mass. [12]

Qvist, B. (1952). Some remarks concerning curves of the second degree in a finite plane. *Ann. Acad. Sci. Fenn.* Ser. A, no. 134. [14]

Ramanujan, S. (1912). Note on a set of simultaneous equations. *J. Indian Math. Soc.* **4**, 94–6. [11]

Rivest, R. L., Shamir, A., and Adleman, L. (1978). A method for obtaining digital signatures and public-key cryptosystems. *Comm. ACM* **21**, 120–6. [16]

Ryser, H. J. (1963). *Combinatorial mathematics*. Carus Monograph 14, Math. Assoc. America. [10]

Schönheim, J. (1968). On linear and non-linear single-error-correcting q-nary perfect codes. *Info. and Control* **12**, 23–6. [9]

Segre, B. (1954). Sulle ovali nei piani lineari finiti. *Atti Accad. Naz. Lincei Rend.* **17**, 1–2. [14]

—— (1955). Curve razionali normali e k-archi negli spazi finiti. *Ann. Mat. Pura Appl.* **39**, 357–79. [15]

—— (1961). *Lectures on modern geometry*. Cremonese, Rome. [15]

Selmer, E. S. (1967). Registration numbers in Norway: some applied number theory and psychology. *Journal of the Royal Statistical Society*, Ser. A **130**, 225–31. [11]

Shannon, C. E. (1948). A mathematical theory of communication. *Bell Syst. Tech. J.* **27**, 379–423 and 623–56. [6]

Singleton, R. C. (1964). Maximum distance q-nary codes. *IEEE Trans. Info. Theory* **10**, 116–18. [10, 15]

Slepian, D. (1960). Some further theory of group codes. *Bell Syst. Tech. J.* **39**, 1219–52. [6]

Sloane, N. J. A. (1981). Error-correcting codes and cryptography. In Klarner, D. A., *The mathematical Gardner*, Wadsworth, Belmont, Calif., pp. 346–382. [16]

—— (1982). Recent bounds for codes, sphere packings and related problems obtained by linear programming and other methods. *Contemporary Mathematics* **9**, 153–85. [2]

Snover, S. L. (1973). The uniqueness of the Nordstrom–Robinson and the Golay binary codes. Ph.D. Thesis, Department of Mathematics, Michigan State Univ. [9]

Stinson, D. R. (1984). A short proof of the nonexistence of a pair of orthogonal Latin squares of order six. *J. Comb. Theory*, Ser. A **36**, 373–76. [9]

Tarry, G. (1901). Le problème des 36 officiers. *C. R. Acad. Sci. Paris* **2**, 170–203. [9]

Thas, J. A. (1968). Normal rational curves and k-arcs in Galois spaces. *Rend. Mat.* **1**, 331–4. [15]

—— (1969). Connection between the Grassmannian $G_{k-1;n}$ and the set of the k-arcs of the Galois space $S_{n,q}$. *Rend. Mat.* **2**, 121–34. [15]

Tietäväinen, A. (1973). On the nonexistence of perfect codes over finite fields. *SIAM J. Appl. Math.* **24**, 88–96. [9]

—— (1980). Bounds for binary codes just outside the Plotkin range. *Info. and Control* **47**, 85–93. [2]

Varshamov, R. R. (1957). Estimate of the number of signals in error correcting codes. *Dokl. Akad. Nauk SSSR* **117**, 739–41. [8]

Vasil'ev, J. L. (1962). On nongroup closepacked codes. *Probl. Kibernat.* **8**, 337–39. (In Russian), translated in *Probleme der Kybernetik* **8** (1965), 375–78. [9]

Verhoeff, J. (1969). *Error detecting decimal codes.* Mathematical Centre Tracts 29, Mathematisch Centrum, Amsterdam. [11]

Verhoeff, T. (1985). Updating a table of bounds on the minimum distance of binary linear codes. Eindhoven University of Technology Report 85-WSK-01. [14]

Weber, E. W. (1983). On the football pool problem for 6 matches: a new upper bound. *J. Comb. Theory*, Ser. A **35**, 106–8. [8]

Zinov'ev, V. A. and Leont'ev, V. K. (1973). The nonexistence of perfect codes over Galois fields. *Problems of Control and Info. Theory* **2**, 123–32. [9]

Index

[*Note:* the bibliography on pages 243–8 serves as a comprehensive index of authors' names since each entry is followed by the numbers of those chapters which refer to that entry.]